Advanced Structures, Fault Diagnosis and Tolerant Control of Permanent Magnet Synchronous Motors

Advanced Structures, Fault Diagnosis and Tolerant Control of Permanent Magnet Synchronous Motors

Editors

Gan Zhang
Hao Hua

MDPI

Basel • Beijing • Wuhan • Barcelona • Belgrade • Novi Sad • Cluj • Manchester

Editors

Gan Zhang
Southeast University
Nanjing
China

Hao Hua
Shanghai Jiao Tong University
Shanghai
China

Editorial Office
MDPI
St. Alban-Anlage 66
4052 Basel, Switzerland

This is a reprint of articles from the Special Issue published online in the open access journal *Energies* (ISSN 1996-1073) (available at: https://www.mdpi.com/journal/energies/special_issues/ASFDTCPMSM).

For citation purposes, cite each article independently as indicated on the article page online and as indicated below:

Lastname, A.A.; Lastname, B.B. Article Title. *Journal Name* **Year**, *Volume Number*, Page Range.

ISBN 978-3-7258-1169-4 (Hbk)
ISBN 978-3-7258-1170-0 (PDF)
doi.org/10.3390/books978-3-7258-1170-0

Contents

energies

MDPI

Article

Design, Analysis, and Optimization of Permanent Magnet Vernier Machines Considering Rotor Losses

Wenjie Wu, Liang Xu * and Bin Liu

School of Electrical and Information Engineering, Jiangsu University, Zhenjiang 212013, China; 2211907021@stmail.ujs.edu.cn (W.W.); 2212107009@stmail.ujs.edu.cn (B.L.)
* Correspondence: xuliang0511@ujs.edu.cn

Abstract: The purpose of this paper is the design, analysis, and optimization of a new structure of a permanent magnet vernier machine (PMVM) with a high torque density and low rotor losses. First, the modulation principle and topology of this PMVM is introduced. Then, its average torque and rotor loss are enhanced and reduced by optimizing the flux modulation poles distribution. For the sake of further reducing the rotor losses on the premise of maintaining the torque density, the contribution of the air gap flux density harmonics to the rotor loss is analyzed. Then, a new topological structure of a rotor with a flux barrier is introduced to reduce the rotor losses due to the decrease of each harmful harmonic. Through the analysis of the structure of the PMVM with the flux barrier, the influence of the parameters on the performance is built. After that, a multi-objective optimization algorithm is used to optimize the PMVM so as to obtain the optimal performance. Moreover, the electromagnetic performance comparison between the newly proposed machine and the original one is presented to indicate that lower rotor losses can be obtained by the proposed machine when the torque is ensured. Finally, a prototype of proposed PMVM is built and further tested to verify the validities of the theoretical and finite-element analyses.

Keywords: permanent magnet vernier machine; rotor losses; vernier; flux barriers; harmonics analysis; finite-element analysis

Citation: Wu, W.; Xu, L.; Liu, B. Design, Analysis, and Optimization of Permanent Magnet Vernier Machines Considering Rotor Losses. *Energies* **2022**, *15*, 1502. https://doi.org/10.3390/en15041502

Academic Editor: Ryszard Palka

Received: 11 January 2022
Accepted: 14 February 2022
Published: 17 February 2022

1. Introduction

Owing to the merits of high efficiency and torque, the permanent magnet (PM) machine has been widely researched and applied in a lot of fields [1–6]. In order to achieve the requirement of a high torque at a low machine size, various high torque density PM machines have been proposed at present, in which a permanent magnet vernier machine (PMVM) as a relatively new machine topology has attracted wide attention because of the characteristics of a simple structure and high torque density [7–14]. In the high torque machine family, PMVM is considered to be a good potential candidate for the field of direct-drive [15–18].

In recent years, plenty of research has been done on PMVM [19]. The working principle of PMVM was intensely analyzed in [20]. Compared with the conventional PM machine, PMVM obtains a larger torque density because of its special working principle of the so-called magnetic gearing effect. In [21], the effect of salient poles and the modulation principle were further analyzed by the theory magnetic gearing effect. On the other hand, in order to improve the electromagnetic performance of PMVM, a lot of new topologies were proposed. In [22], an advanced PMVM with multiple working harmonics was proposed, where new working harmonics are modulated by introducing the structure of flux modulation poles (FMPs) and thus improving its torque density. In [23–26], by adopting the doubly PM excited structure in the PMVMs, the amplitudes of multiple working harmonics were further improved. In addition, their torque densities were effectively boosted.

At the same time, there has been extensive research about the optimization of the machines in recent years. The various optimization technologies and the development trend of the machine for its structural size, shape, and topology designs have been discussed in [27]. Moreover, in order to avoid the negative effects of parameters and material tolerances on performance, a robust optimization was utilized in the machine to improve the reliability of the performance. In [28], different designs of experiment methodologies were integrated into encapsuled FEM models using a Digital-Twin-Distiller. Through those methodologies, the tolerances of the machine were analyzed in the process of optimization with a reduction in the computational time. In [29], a PMVM was optimized considering the effect of the manufacturing tolerance. Through the deterministic optimization algorithm and design for the six sigma method, the power factor, torque, and robustness of the PMVM were improved simultaneously.

However, some problems in the design, analysis, and optimization of PMVM need to be solved. Especially noteworthy, abundant harmonics in PMVM can provide a high torque while causing a large electromagnetic loss. Moreover, harmonics with rotation speed and direction have a greatly significant effect on rotor loss [30]. The rotor losses will produce an additional overheat of machine, which may result in a reduction of machine performances and even irreversible demagnetization of PMs. Therefore, rotor loss suppression has become an important direction for developing PMVM. Nowadays, the rotor loss suppression for the conventional PM machine is mainly carried out from the following aspects, such as optimizing the winding structure, segmenting PMs, and changing the stator and rotor structure. In [31], a multilayer winding configuration was adopted for a conventional interior PM machine to reduce stator magnetomotive force (MMF) harmonics, and then the rotor losses could be reduced. Meanwhile, segmenting the PMs is a common method to reduce PM eddy current (EC) loss [32]. However, up until now, the loss analysis and suppression of PMVM was often neglected in previous work. In addition, except for PM segmentation, other rotor loss suppression approaches for the conventional PM machine could not be applied to the PMVM directly. This is likely to be invalid with suppression of the harmonics producing rotor loss due to the multiple working harmonics principle. Consequently, the torque is also significantly reduced with the loss reduction.

In order to solve this problem, the structure of the flux barriers was designed based on the original PMVM. With the desirable suppression effect to the harmful harmonic, the flux barrier design is facilitated to reduce the rotor loss significantly and maintain the torque. Of course, the introduction of the flux barrier structure increases the structural complexity to some extent. In this paper, through the analysis of rotor loss and torque, key parameters of flux barrier and FMP structures on the performance can be determined. Then, a multi-objective optimization algorithm is adopted for PMVM with flux barriers so as to achieve both a high torque density and low rotor loss.

The rest of this paper is organized as follows. The topology of the original PMVM and the modulation principle of the PMVM are introduced in Section 2. The torque of the PMVM is improved by optimizing the FMPs distribution in Section 3. For the sake of further reducing the rotor losses on the premise of maintaining the torque density, the loss component of the PMVM is analyzed in Section 4. Based on this process, a novel topological structure of the rotor is introduced to reduce the rotor losses in Section 5. The function of flux barriers on the armature reaction field is analyzed using a simple analytical model. Then, the flux barriers are designed considering the width and depth. In Section 6, the PMVM with a flux barrier is optimized using an optimization algorithm and is compared with the original PMVM. In Section 7, the electromagnetic performances of the original and proposed PMVMs, such as flux linkage, air-gap flux density, no-load back-EMFs, rotor losses, and torque, are analyzed using the method of finite element analysis (FEA). Section 8 presents the experiments on the prototypes of both machines. Lastly, the conclusions of this paper are made in Section 9.

2. Topology and Working Principle

Figure 1 presents the topological structure of the both PMVMs. It can be seen that the machine adopts an outer-rotor structure with 31 pole pairs of PMs evenly distributed on the rotor yoke. There are 20 stator teeth in the stator, and each stator tooth splits into two FMPs. The modulation functions of these FMPs are same as the ferromagnetic pieces in the magnetic gear. Moreover, the winding pattern of the machine adopts the concentrated winding, which effectively reduces the end length of the windings. In addition, compared with the original PMVM structure, the rotor loss of the PMVM is suppressed by the proposed PMVM with the new structure of flux barriers at the rotor yoke.

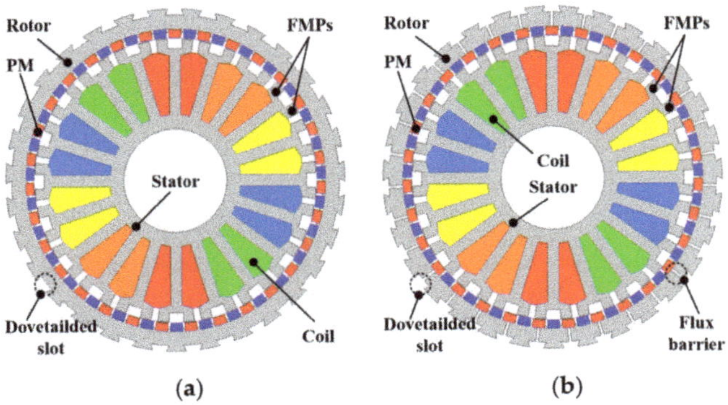

Figure 1. Topology of the original PMVM: (**a**) original and (**b**) proposed.

The operational principle of PMVM is based on the magnetic gearing effect. According to the theory of the magnetic field modulation effect, a new harmonic component is modulated to match the low-order armature harmonic with the high-order PM harmonic. In this way, the magnetic field can change significantly due to the subtle change of the position caused by the rotor rotation, which achieves a high torque density at a low speed. To improve the torque performance by making use of the magnetic gearing effect, the calculation formula between the pole pairs of the armature winding P_a and pole pairs of the rotor P_r and FMPs P_f can be expressed as follows:

$$P_a = P_f - P_r \tag{1}$$

3. Design and Optimization of FMPs

It is worth mentioning that, currently, most PMVMs use uniform distributed FMPs. However, FMPs with different distributions will also have a certain influence on the performance of machine. Figure 2 shows the detailed design parameters for the FMPs, where θ_1 (FMP-slot width) represents the width between two FMPs on the same stator tooth and θ_2 (FMP width) represents the width of a single FMP. Then, the FMP slot width ratio k_s and FMP width ratio k_p can be expressed as follows:

$$k_s = \frac{\theta_1 P_f}{\pi} \text{ and } k_p = \frac{\theta_2 P_f}{\pi} \tag{2}$$

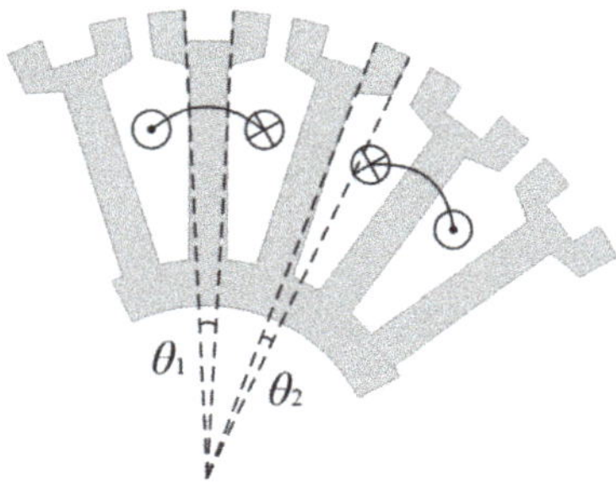

Figure 2. Detailed design parameters for FMPs.

When $k_s + k_p = 2$, all the FMP-slot widths and slot opening widths are equal, i.e., uniformly distributed FMPs. Conversely, when $k_s + k_p \neq 2$, the width between FMPs will change and become nonuniformly distributed FMPs. With the change of k_s and k_p, the air-gap field of the machine will change accordingly, which will affect the performance of the machine. Figures 3 and 4 illustrate the average torque and rotor loss variations of the PMVM with k_s and k_p. As k_s and k_p increase, the average torque increases at the beginning. However, the average torque begins to decrease when k_s and k_p exceed a certain range. The average torque is improved by 21.4% at the maximum value compared with the fundamental model ($k_s = k_p = 1$). On the other hand, with the increase of k_s and k_p, the rotor losses present a trend of first increasing and then decreasing. When the average torque reaches its maximum value, the rotor losses decrease by about 10%. Although the optimized PMVM reduces some of the rotor losses by optimizing the FMPs, the remaining rotor losses cannot be further reduced. To ensure the torque density and minimize the losses of PMVM, the factors affecting the loss need to be further investigated and analyzed.

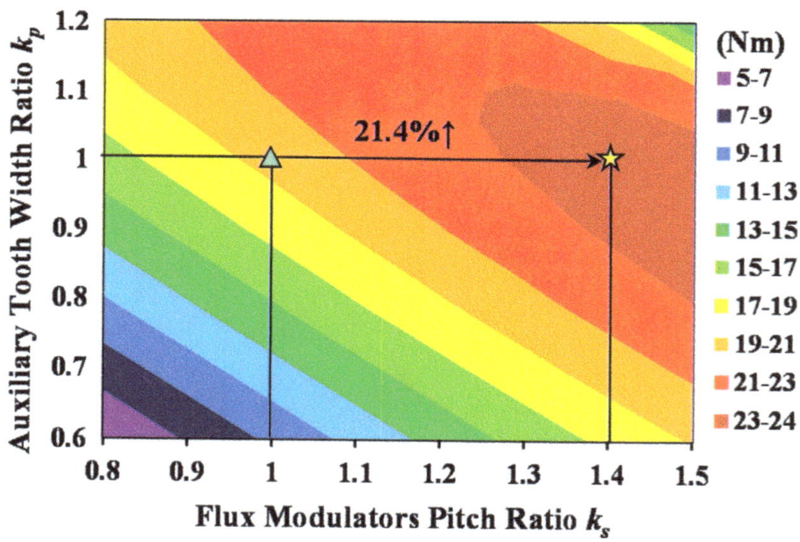

Figure 3. Variation of average torque with k_s and k_p in PMVM.

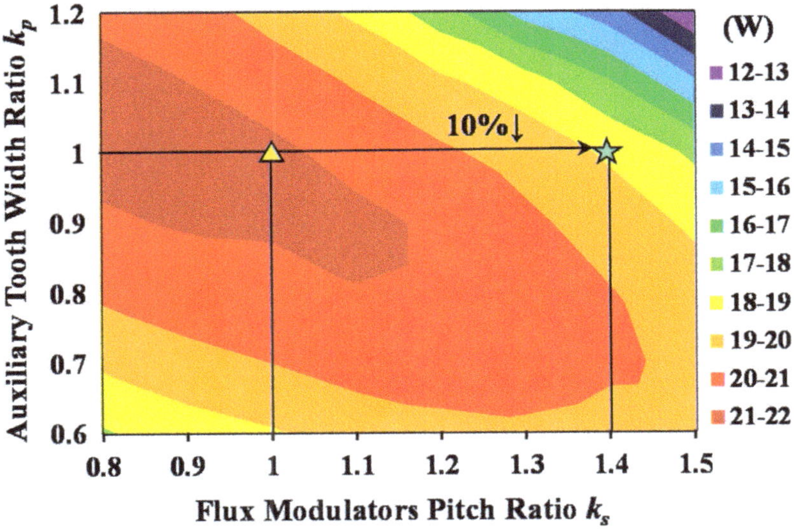

Figure 4. Variation of rotor losses with k_s and k_p in PMVM.

4. Rotor Losses Analysis

It should be noted that after field modulation, a large number of harmonics appear in the PMVM. This could produce large rotor losses, because these harmonics are asynchronous because of the rotation of the rotor. Furthermore, these negative effects are more obvious due to the increase in the armature reaction field. Hence, it is important to analyze the loss of PMVM.

The relationship between EC loss and the parameters of PMs was expressed in [33,34]. The EC loss with regard to the PMs can be achieved as follows:

$$P_{eddy} = abd \cdot \frac{\omega_k^2 B_{PMk}^2 \sigma b^2}{24}\left(1 - \frac{192}{\pi^5} \cdot \frac{b}{a} \cdot \sum_{n=0}^{+\infty} \frac{\tanh\left(\frac{(2n+1)\pi a}{2b}\right)}{(2n+1)^5}\right) \tag{3}$$

where a and b are the height and weight of the PMs, respectively. Moreover, d is the thickness of the PMs. σ and ω_k indicate the PM conductivity and the k-th harmonic rotation speed, respectively. B_{PMk} is the amplitude of the k-th harmonic PM flux density. From Equation (3), it can be seen that not only the structural parameters of PMs, but also the harmonic amplitudes of the flux density had a great influence on PM EC loss.

Core loss, which consists of EC loss and hysteresis loss, can be approximately expressed from the flux density of each harmonic, as follows:

$$P_{Core} = \sum_k \left(A_e f_k^2 B_{Corek}^2 + A_h f_k B_{Corek}^2\right) \tag{4}$$

where A_e and f_k are the EC loss coefficient and the frequency of different order harmonics, respectively. A_h and B_{Corek} are the coefficient of the hysteresis loss and the k order harmonic amplitude of rotor core flux density, respectively. From Equation (4), it can be seen that the frequency and flux density of each harmonic mainly affect core loss.

Because of the principle of the modulation effect, the harmonic components generated by PMs can be expressed as follows

$$P_{pm}(i,j) = |iP_r \pm jN_s| \tag{5}$$

where $i = 1, 3, 5, \ldots, j = 0, 1, 2, \ldots$. It can be inferred that the working harmonics of the machine are the 9th, 11th, 29th, 31st, etc. Figures 5 and 6 show the no-load and armature reaction air-gap flux density, respectively. This reveals that the harmonic components generated by the PM field were 9th, 11th, 29th, and 31st orders, while in addition to these harmonics, the armature reaction field also had 1st, 19th, and 21st order harmonics. Meanwhile, the 31st harmonic which mainly caused by the fundamental MMF move synchronously with the rotor, which is a static harmonic and does not produce losses. Figure 7 shows the on-load machine rotor losses of the original PMVM, including PM EC loss and rotor core loss under the on-load situation. This indicates that the PM EC loss corresponding to the 9th harmonic reached the peak value, while 1st harmonic was the second largest value. The PM EC loss with regard to the 1st harmonic occupied about 22% of the whole PM EC loss. In addition, the rotor core loss was mainly determined by the 1st harmonic, which occupied nearly 96% of the total rotor core loss. In summary, it was found that the 1st harmonic has a huge influence on rotor losses. On the other hand, the 1st harmonic was found to be a non-working harmonic that has no effect on torque. Therefore, through reducing the 1st harmonic, the rotor losses could be reduced while maintaining the torque density.

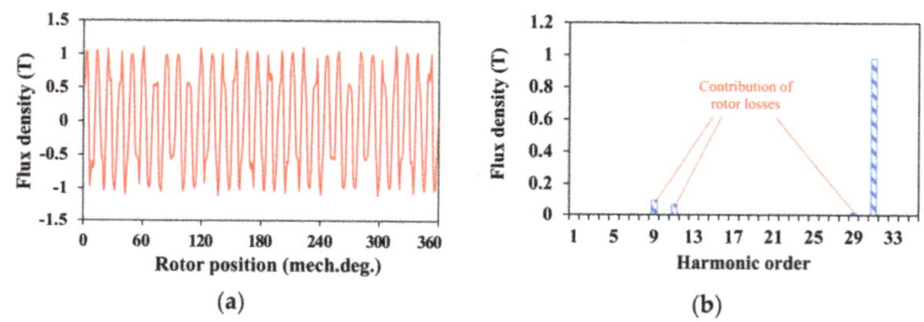

Figure 5. No-load flux density harmonics distribution of PMVM: (**a**) waveform and (**b**) spectrum.

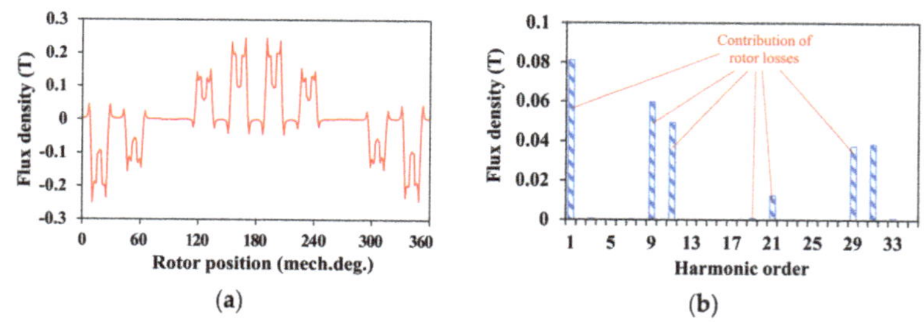

Figure 6. Armature reaction flux density harmonics distribution of PMVM: (**a**) waveform and (**b**) spectrum.

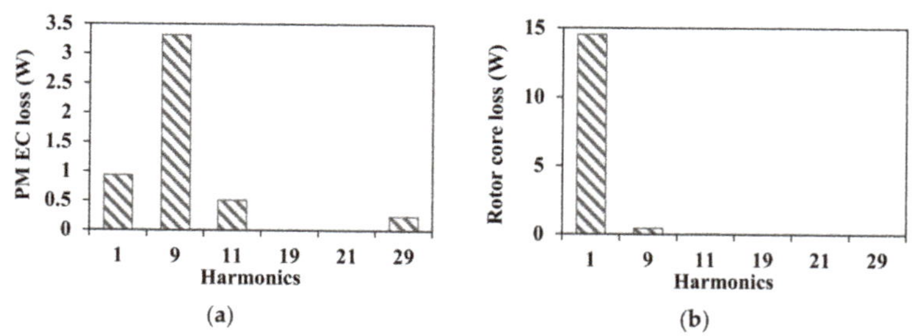

Figure 7. The result of rotor losses in the original PMVM: (**a**) PM EC loss and (**b**) rotor core loss.

5. Low Rotor Loss Design

In order to effectively depress the 1st armature harmonic, a new rotor structure of flux barriers is introduced in PMVM. Flux barriers with a suitable size and position are placed at the rotor yoke. Figure 8 shows the detailed design parameters for the flux barriers, where w_{fb} represents the width of a single flux barrier and d_{fb} represents the depth of a single flux barrier.

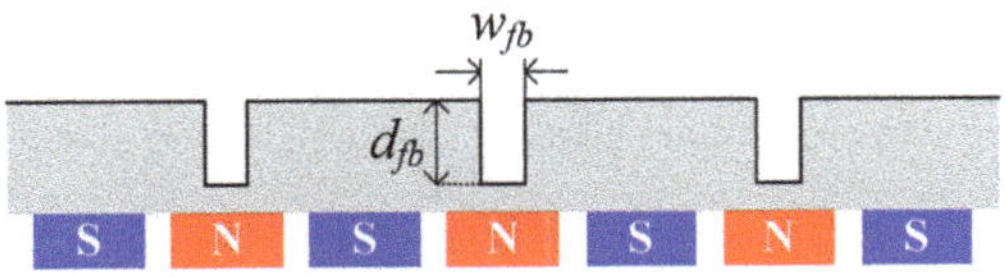

Figure 8. Detailed design parameters for flux barriers.

5.1. Analysis of Armature Reaction Field

The armature reaction field is analyzed by a simple analysis model, as shown in Figure 9.

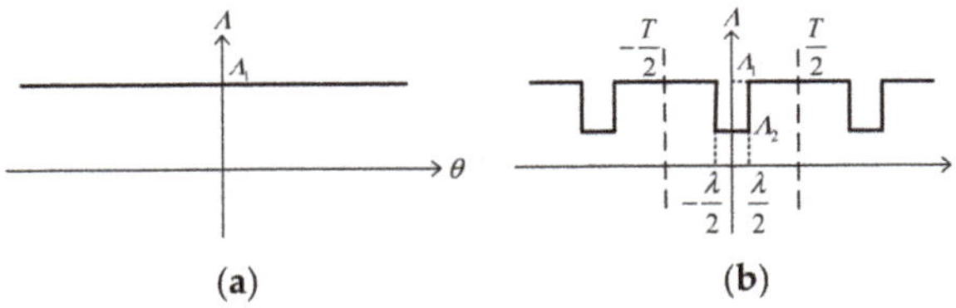

Figure 9. Simple rotor permeance model for the armature reaction field analysis: (**a**) original and (**b**) proposed.

Accordingly, the rotor permeance of the original PMVM and the proposed one can be expressed as

$$\Lambda_{or}(\theta) = \Lambda_1 \tag{6}$$

$$\Lambda_{pr}(\theta) = \Lambda_1 - (\Lambda_1 - \Lambda_2)\frac{\lambda}{T} - \sum_{m=1}^{\infty} \frac{2(\Lambda_1 - \Lambda_2)}{m\pi} \sin\left(\frac{m\pi\lambda}{T}\right) \cos(mP_r\theta) \tag{7}$$

where Λ_1 and Λ_2 are the Fourier coefficients, and λ/T is ratio of flux barrier thickness with regard to the rotor yoke thickness. It is worth noting that the abbreviations of *or* and *pr* represent the original and proposed machines, respectively.

The armature reaction MMF of the PMVM can be expressed as

$$F_{aq}(\theta, t) = \frac{10F_q N I_{\max}}{q\pi} \sum_q \cos(q\theta \pm \omega t) \tag{8}$$

where F_q is the Fourier coefficient, N is the number of turns per phase, $I_{\max}$ is the maximum value of armature current, and $q = 10r \pm 1, r = 0, 1, 2 \ldots$.

Then, through multiplying the armature reaction MMF and original rotor permeance, the original flux density in the armature flux field can be obtained as follows:

$$B_{or}(\theta, t) = \frac{10F_q N I_{\max}}{q\pi} \Lambda_1 \sum_q \cos(q\theta \pm \omega t) \tag{9}$$

Meanwhile, the proposed flux density of the armature reaction can be expressed as

$$\begin{aligned} B_{pr}(\theta,t) &= \frac{10F_q N I_{\max}}{q\pi}\left[\Lambda_1-(\Lambda_1-\Lambda_2)\frac{\lambda}{T}\right]\sum_q \cos(q\theta\pm\omega t) \\ &-\frac{10F_q N I_{\max}}{q\pi^2}(\Lambda_1-\Lambda_2)\sin\left(\frac{m\pi\lambda}{T}\right)\sum_q\sum_{m=1}^{\infty}\cos[(q+mP_r)\theta\pm\omega t] \\ &-\frac{10F_q N I_{\max}}{q\pi^2}(\Lambda_1-\Lambda_2)\sin\left(\frac{m\pi\lambda}{T}\right)\sum_q\sum_{m=1}^{\infty}\cos[(q-mP_r)\theta\pm\omega t] \end{aligned} \quad (10)$$

Comparing Equations (9) and (10), it can be found that the amplitude of the armature reaction flux density will decrease due to the effect of the flux barrier, hence suppressing the rotor loss of the machine to some extent. On the other hand, due to the function of field modulation by flux barrier, $q + mP_r$ and $q - mP_r$ order harmonics are introduced. Figure 10 shows the amplitude variation of q order armature harmonic corresponding to the change of the flux barrier width. From this figure, the amplitude of the 1st harmonic significantly decreases because of the increase in the width of the flux barriers. According to Equations (3) and (4), the loss caused by the 1st harmonic will be greatly reduced. On the other hand, the 31st working harmonic mainly affecting the torque component nearly has no change. Therefore, the torque of PMVM can be maintained and the rotor losses can be lowered.

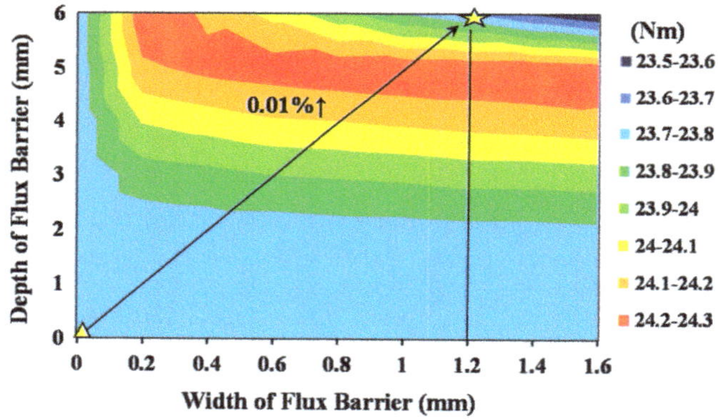

Figure 10. Variation of average torque with w_{fb} and d_{fb} in the proposed PMVM.

5.2. Design of Flux Barriers

The armature reaction will produce some useless low order harmonics, which increase the losses of the machine. These low order harmonics, especially the 1st harmonic produced by the armature reaction, can be effectively suppressed by adopting flux barrier design. Meanwhile, it will not have a significant influence on high order working harmonics, which ensures the torque density. In order to maintain torque density and reduce rotor losses as much as possible, it is very important to choose appropriate flux barriers. Figures 10 and 11 illustrate the average torque and rotor loss variations of PMVM with the flux barrier width w_{fb} and flux barrier depth d_{fb}. As w_{fb} and d_{fb} increase, the average torque also increases at the beginning. Then, the average torque begins to decrease when w_{fb} and d_{fb} exceed a certain range. On the other hand, with the increase in w_{fb} and d_{fb}, the rotor losses are significantly reduced.

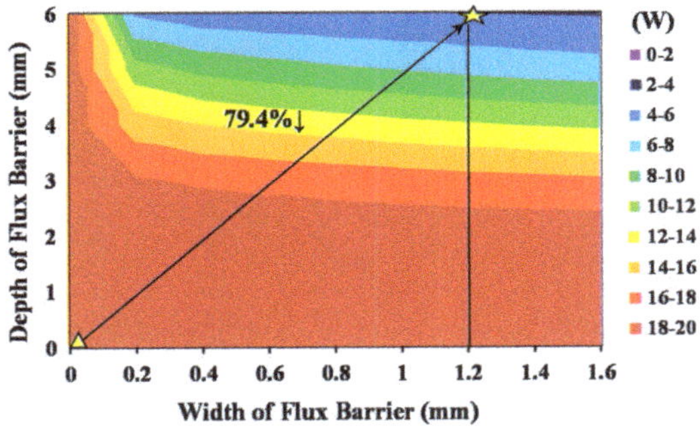

Figure 11. Variation of rotor losses with w_{fb} and d_{fb} in the proposed PMVM.

6. Performance Optimization

Through the above-mentioned design and analysis of the PMVM structure, it can be seen that the FMP slot width ratio k_s, FMP width ratio k_p, the flux barrier width w_{fb}, and the flux barrier depth d_{fb} influenced the performance of PMVM. In order to obtain the optimal torque and rotor loss, a multi-objective optimization non-dominated sorting genetic algorithm II (NSGA-II) was used to optimize the torque and rotor loss. Figure 12 shows the Pareto frontiers of PMVM with flux barriers and original PMVM using this optimization method. It can be seen that all the Pareto design points with regard to the PMVM with flux barrier present lower rotor loss than the original PMVM. In order to obtain minimal rotor loss, the green points were selected as the optimization results to conduct further research. The torque and rotor loss of the selected optimal PMVM with flux barriers were 23.72 Nm and 0.6 W, respectively. Therefore, the parameters k_s and k_p were finally chosen as 1.4 and 1.0, respectively, and w_{fb} and d_{fb} were finally determined to be 1.2 mm and 6 mm, respectively. Compared with the 23.69 Nm torque and 15 W rotor loss of the original machine, the rotor losses of the proposed machine with flux barriers were reduced significantly, while the torque was basically unchanged.

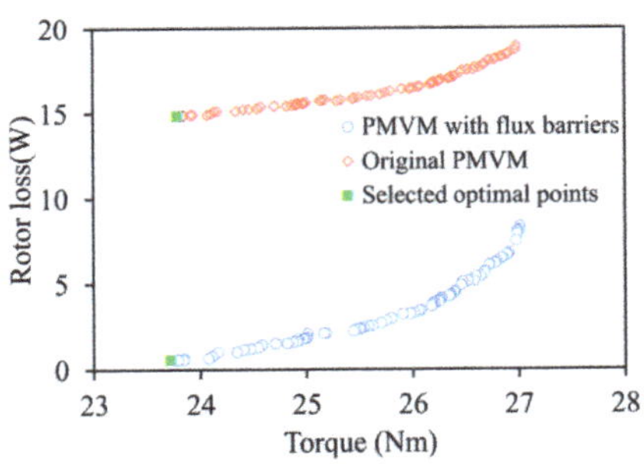

Figure 12. Pareto design frontiers calculated by NSGA-II.

7. Performance Evaluation

The electromagnetic performances of both PMVMs are compared comprehensively by FEA in this section. The results reveal that lower rotor losses could be obtained by the proposed machine without sacrificing the torque density, which verifies the above theoretical analysis.

7.1. Open-Circuit Characteristics

Figure 13 presents the no-load magnetic field distributions of the original PMVM and proposed PMVM. It can be seen that the magnetic field distributions were almost same between the two machines. This phenomenon means that the PM magnetic field was not influenced by flux barriers with an appropriate position and size. The no-load back-EMF waveforms are illustrated in Figure 14. It can be observed that the flux barrier design had little effect on the no-load back-EMFs.

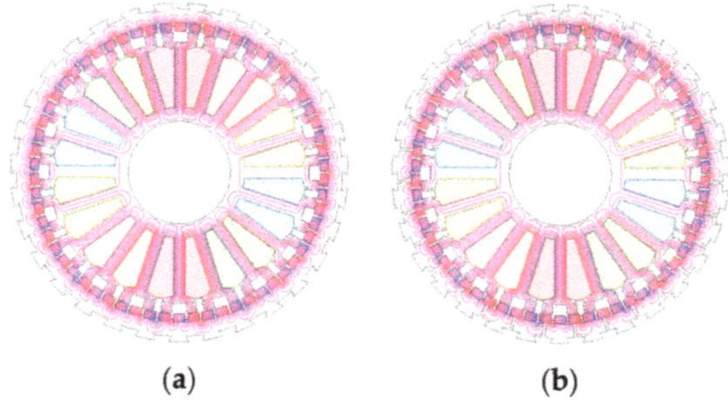

Figure 13. Magnetic field distributions of both PMVMs in open-circuit operation: (**a**) original and (**b**) proposed.

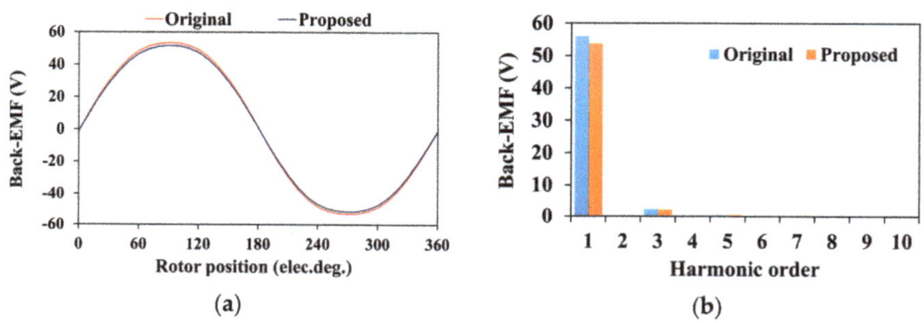

Figure 14. Phase open-circuit back-EMFs of both PMVMs: (**a**) waveform and (**b**) spectrum.

Figure 15 makes a comparison between the no-load flux density waveform and harmonic spectrum of the PMVMs. It can be seen that the waveform and spectrum of the air-gap flux density of the proposed machine were basically coincident with the original one. The new rotor topology had only a slight influence on the PM field, which indicates that the torque performance of the proposed machine will not degrade.

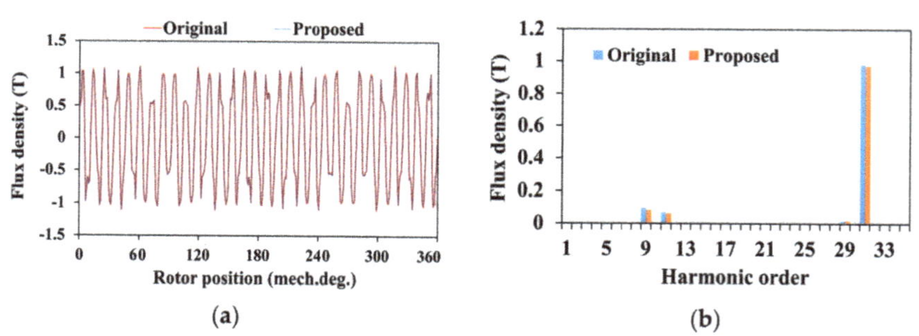

Figure 15. Comparison of the open-circuit flux density of both PMVMs: (**a**) waveform and (**b**) spectrum.

7.2. Armature Reaction Characteristics

Figure 16 depicts the armature reaction magnetic field nephogram. As shown, the proposed machine obtained greater strength for the armature magnetic field than the original one. Figure 17 shows the flux density waveforms of the armature reaction and harmonics of the two PMVMs. As can be seen, the 1st harmonic had a significant decline, while the higher working harmonics were not greatly affected. The obvious weakening of the 1st harmonic from the magnetic field distribution can be seen in Figure 16b. At the same time, some new modulation harmonics could be produced by the flux barriers (such as the 20th and 22th order). It is worth mentioning that if the core saturation was high, the torque of the PMVM would be limited. Through the new proposed flux barrier in the rotor, the core saturation could be significantly reduced, which increased the torque capacity of the machine. As a result, although the back-EMF of the machine shown in Figure 14 decreased slightly, the torque of the machine could improve to a certain extent.

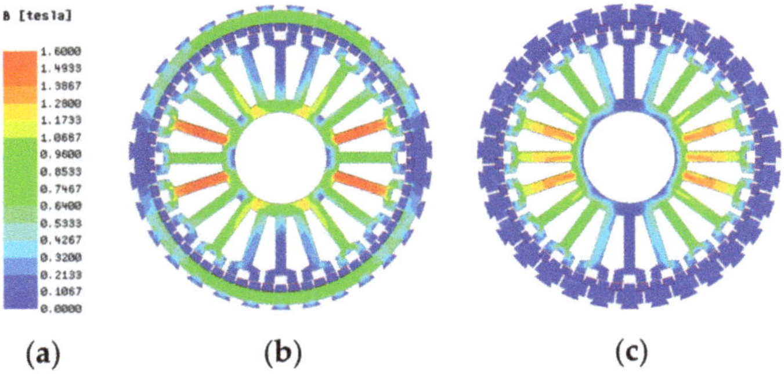

Figure 16. Comparison of the armature reaction field nephogram: (**a**) color scale, (**b**) original, and (**c**) proposed.

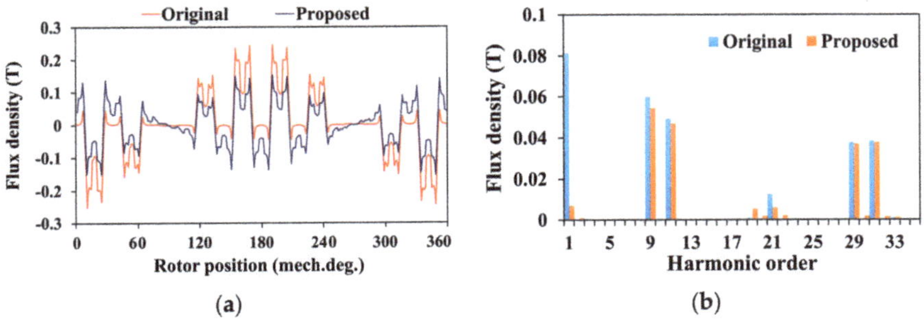

Figure 17. Comparison of the flux density of two PMVMs in the armature reaction: (**a**) waveform and (**b**) scheme.

7.3. Rotor Losses

Figure 18 shows the rotor losses by FEA. From Figure 18, the proposed machine obtained much lower rotor losses than the original machine. The PM EC loss decreased by 26% while the rotor core loss of the proposed machine was also reduced by 96%. Figure 19 depicts the variation of the rotor losses versus rotor speed of both machines. As observed, with increasing the rotor speed, the increase of rotor core loss of the original machine was much higher than that of the proposed one. Meanwhile, with increasing the rotor speed, the increment of PM EC loss of the original PMVM was larger than that of the proposed one. Thus, the proposed machine could offer a much lower rotor loss when increasing the rotor speed.

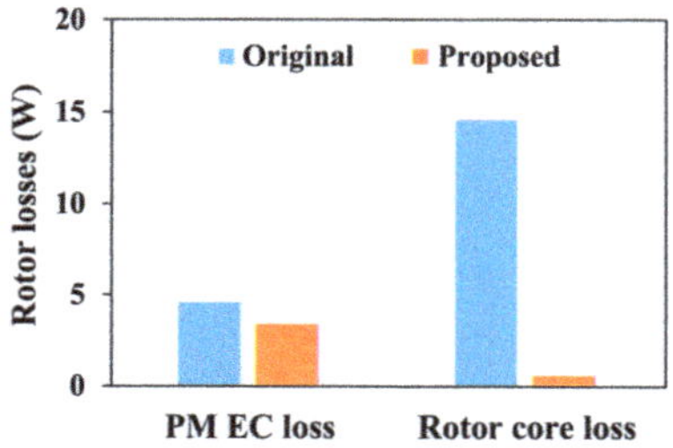

Figure 18. Comparison of the rotor losses between the original and proposed machines.

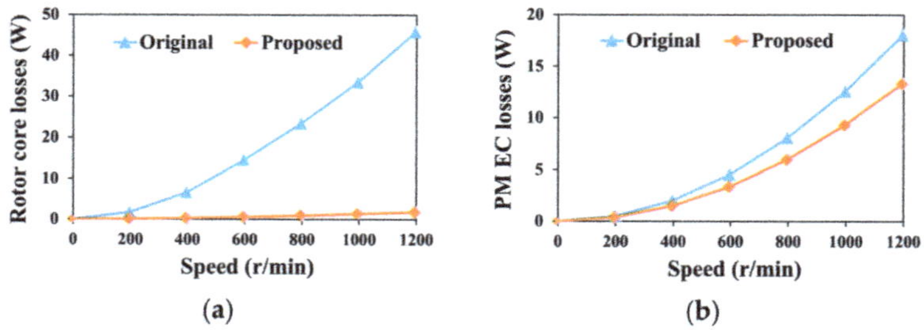

Figure 19. Rotor losses with the rotor speed of both machines: (**a**) rotor core loss and (**b**) PM EC loss.

7.4. Torque Capacity

The average torque and torque pulsation are shown in Figure 20. It can be seen that both the original and proposed PMVM had a high torque and low torque ripple, which is an advantage of PMVM. Meanwhile, the average torque and torque ripple of both PMVMs was basically unchanged. These results prove the correctness of the preceding analysis.

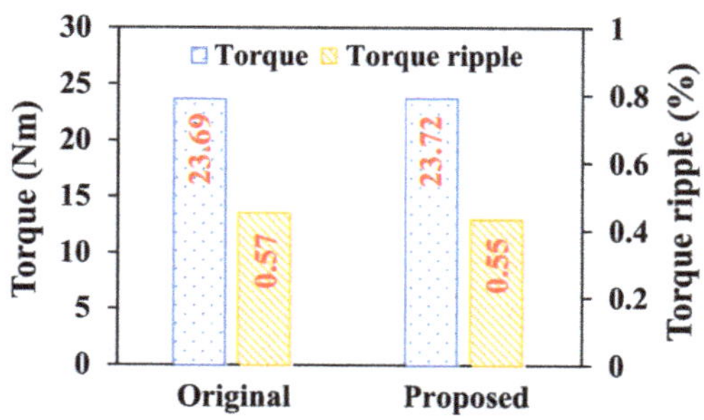

Figure 20. Average torque and torque pulsation of both PMVMs.

8. Experimental Validation

Based on the above analysis, a new PMVM with low rotor losses was built. The new proposed PMVM could significantly reduce rotor losses, while maintaining the torque density. To verify the effectiveness of the analysis results through the experiment, the original and proposed PMVMs were manufactured and tested, respectively. The prototypes for both PMVMs were manufactured as shown in Figure 21.

(a) (b) (c)

Figure 21. Prototype machines: (**a**) stator, (**b**) original rotor, and (**c**) new rotor with flux barriers.

Throughout the measurements, the no-load back-EMFs of the original and proposed machines were tested at 600 r/min, as shown in Figure 22. It can be seen that the differences between the measured and FEA results of the original and proposed PMVMs were lower than 6%. The differences were mainly due to experimental errors. In addition, the amplitude of the no-load back-EMF of both machines was basically unchanged, and the waveforms were approximately sinusoidal. This also verified the consistency of the machine performance between the simulation and experiment by introducing the flux barriers in the proposed machine.

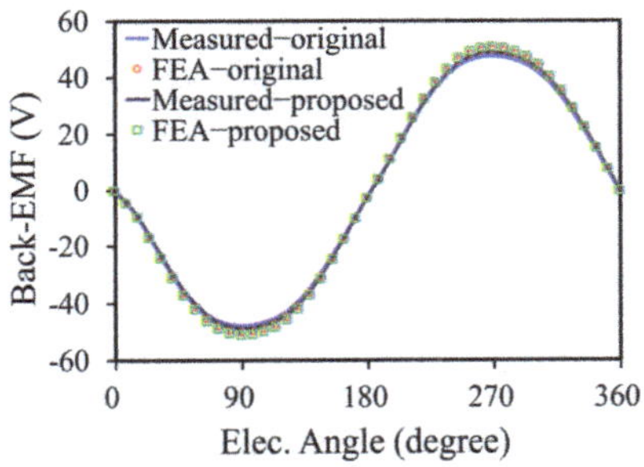

Figure 22. FEA and measured back EMFs of the original and proposed PMVMs.

9. Conclusions

In this paper, a new PMVM with a high torque density and low rotor losses is proposed. First of all, this paper introduces the topological structure and working principle of PMVM. Then, the torque of PMVM is effectively enhanced by optimizing the parameters of FMPs. To reduce rotor losses, the loss component of PMVM is analyzed. The results show that the first harmonic causes rotor losses while having no effect on the average torque. Therefore, the 1st harmonic is considered to be reduced in order to maintain the torque and reduce the rotor loss. Based on this process, a new rotor structure with flux barriers is designed. The influence of the flux barriers on the first harmonic is revealed. It is found that with appropriate flux barrier design, the first harmonic produced by the armature reaction field can be greatly reduced and the reduction of working harmonics is slight. As a result, the rotor losses are effectively reduced while maintaining the torque density. Furthermore, the proposed and original PMVMs are compared by FEA. Finally, the two PMVMs are fabricated and final test results verify the effectives of the theoretical analysis and FEA results.

Author Contributions: Conceptualization, L.X.; manufacturing, W.W. and L.X.; software, W.W. and B.L.; validation; B.L. and L.X.; writing—original draft preparation, W.W.; writing—review and editing, L.X.; supervision, L.X. All authors have read and agreed to the published version of the manuscript.

Funding: This work was supported in part by the National Natural Science Foundation of China (52177044), the Hong Kong Scholars Program (XJ2019031), the Natural Science Foundation of Jiangsu Higher Education Institutions (21KJA470004), and the Priority Academic Program Development of Jiangsu Higher Education Institutions.

Institutional Review Board Statement: Not applicable.

Informed Consent Statement: Not applicable.

Data Availability Statement: Not applicable.

Conflicts of Interest: The authors declare no conflict of interest.

References

1. Li, X.; Xue, Z.; Yan, X.; Zhang, L.; Ma, W.; Hua, W. Low-complexity multivector-based model predictive torque control for PMSM with voltage preselection. *IEEE Trans. Power Electron.* **2021**, *36*, 11726–11738. [CrossRef]
2. Wang, B.; Wang, J.; Griffo, A.; Sen, B. Experimental Assessments of a Triple Redundant Nine-Phase Fault-Tolerant PMA SynRM Drive. *IEEE Trans. Ind. Electron.* **2019**, *66*, 772–783. [CrossRef]
3. Li, X.; Xue, Z.; Zhang, L.; Hua, W. A low-complexity three-vector-based model predictive torque control for SPMSM. *IEEE Trans. Power Electron.* **2021**, *36*, 13002–13012. [CrossRef]
4. Chen, Q.; Xu, G.; Zhai, F.; Liu, G. A novel spoke-type PM motor with auxiliary salient poles for low torque pulsation. *IEEE Trans. Ind. Electron.* **2020**, *67*, 4762–4773. [CrossRef]
5. Jiang, T.; Zhao, W.; Xu, L.; Ji, J. A novel parallel hybrid excitation field modulated machine with efficient utilization of multiworking harmonics. *IEEE Trans. Ind. Electron.* **2020**, *69*, 1177–1188. [CrossRef]
6. Zhang, G.; Hua, W.; Han, P. Quantitative Evaluation of the Topologies and Electromagnetic Performances of Dual-Three-Phase Flux-Switching Machines. *IEEE Trans. Ind. Electron.* **2018**, *65*, 9157–9167. [CrossRef]
7. Xu, L.; Liu, G.; Zhao, W.; Yang, X.; Cheng, R. Hybrid stator design of fault-tolerant permanent-magnet vernier machines for direct-drive applications. *IEEE Trans. Ind. Electron.* **2017**, *64*, 179–190. [CrossRef]
8. Zhao, X.; Niu, S. Design of a novel parallel-hybrid-excited vernier reluctance machine with improved utilization of redundant winding harmonics. *IEEE Trans. Ind. Electron.* **2018**, *65*, 9056–9067. [CrossRef]
9. Jiang, S.; Liu, G.; Zhao, W.; Xu, L.; Chen, Q. Modeling and analysis of spoke-type permanent magnet vernier machine based on equivalent magnetic network method. *Chin. J. Elect. Eng.* **2018**, *4*, 96–103.
10. Shi, Y.; Jian, L. A novel dual-permanent-magnet-excited machine with flux strengthening effect for low-speed large-torque applications. *Energies* **2018**, *11*, 158. [CrossRef]
11. Jang, D.; Chang, J. Investigation of doubly salient structure for permanent magnet vernier machines using flux modulation effects. *IEEE Trans. Energy Convers.* **2019**, *34*, 2019–2028. [CrossRef]
12. Xu, L.; Wu, W.; Zhao, W. Airgap magnetic field harmonic synergetic optimization approach for power factor improvement of PM vernier machines. *IEEE Trans. Ind. Electron.* [CrossRef]
13. Du, K.; Xu, L.; Zhao, W.; Liu, G. Analysis and Design of a Fault-Tolerant Permanent Magnet Vernier Machine With Improved Power Factor. *IEEE Trans. Ind. Electron.* **2022**, *69*, 4353–4363. [CrossRef]
14. Li, X.; Chau, K.; Cheng, M. Analysis design and experimental verification of a field-modulated permanent-magnet machine for direct-drive wind turbines. *IET Electr. Power Appl.* **2015**, *9*, 150–159. [CrossRef]
15. Xu, L.; Zhao, W.; Liu, G.; Song, C. Design optimization of a spoke-type permanent-magnet vernier machine for torque density and power factor improvement. *IEEE Trans. Veh. Technol.* **2019**, *68*, 3446–3456. [CrossRef]
16. Tlali, P.; Wang, R.; Gerber, S.; Botha, C.; Kamper, M. Design and performance comparison of vernier and conventional PM synchronous wind generators. *IEEE Trans. Ind. Appl.* **2020**, *56*, 2570–2579. [CrossRef]
17. Raihan, M.; Baker, N.; Smith, K.; Almoraya, A. Development and testing of a novel cylindrical permanent magnet linear generator. *IEEE Trans. Ind. Appl.* **2020**, *56*, 3668–3678. [CrossRef]
18. Kwon, J.; Kwon, B. Investigation of dual-stator spoke-type vernier machine for EV application. *IEEE Trans. Magn.* **2018**, *54*, 1–5. [CrossRef]
19. Wu, F.; Refaie, A. Permanent magnet vernier machines: A review. *IET Electr. Power Appl.* **2019**, *13*, 127–137. [CrossRef]
20. Kim, B.; Lipo, T. Operation and design principles of a PM vernier motor. *IEEE Trans. Ind. Appl.* **2014**, *50*, 3656–3663. [CrossRef]
21. Cheng, M.; Wei, H.; Han, P.; Feng, X. Analysis of airgap field modulation principle of simple salient poles. *IEEE Trans. Ind. Electron.* **2019**, *66*, 2628–2638. [CrossRef]
22. Zou, T.; Li, D.; Qu, R.; Jiang, D.; Li, J. Advanced high torque density PM vernier machine with multiple working harmonics. *IEEE Trans. Ind. Appl.* **2017**, *53*, 5295–5304. [CrossRef]
23. Wang, Q.; Niu, S.; Yang, L. Design optimization and comparative study of novel dual-PM excited machines. *IEEE Trans. Ind. Electron.* **2017**, *64*, 9924–9933. [CrossRef]
24. Lin, Q.; Niu, S.; Fu, W. Design and optimization of a dual-permanent-magnet vernier machine with a novel optimization model. *IEEE Trans. Magn.* **2020**, *6*, 1–5. [CrossRef]
25. Xu, L.; Zhao, W.; Liu, G.; Ji, J.; Niu, S. A novel dual-permanent-magnet-excited machine with non-uniformly distributed permanent-magnets and flux modulation poles on the stator. *IEEE Trans. Veh. Technol.* **2020**, *69*, 7104–7115. [CrossRef]
26. Gao, Y.; Doppelbauer, M.; Qu, R.; Li, D.; Ding, H. Synthesis of a flux modulation machine with permanent magnets on both stator and rotor. *IEEE Trans. Ind. Appl.* **2021**, *57*, 294–305. [CrossRef]

27. Bramerdorfer, G.; Tapia, J.; Pyrhonen, J.; Cavagnino, A. Modern Electrical Machine Design Optimization: Techniques, Trends, and Best Practices. *IEEE Trans. Ind. Electron.* **2018**, *65*, 7672–7684. [CrossRef]
28. Orosz, T.; Gadó, K.; Katona, M.; Rassõlkin, A. Automatic Tolerance Analysis of Permanent Magnet Machines with Encapsuled FEM Models Using Digital-Twin-Distiller. *Processes* **2021**, *9*, 2077.
29. Xu, L.; Wu, W.; Zhao, W.; Liu, G.; Niu, S. Robust design and optimization for a permanent magnet vernier machine with hy-brid stator. *IEEE Trans. Energy Convers.* **2020**, *9*, 2086–2094. [CrossRef]
30. Xu, L.; Zhao, W.; Wu, M.; Ji, J. Investigation of slot–pole combination of dual-permanent-magnet-excited vernier machines by using air-gap field modulation theory. *IEEE Trans. Transport. Electrif.* **2019**, *5*, 1360–1369. [CrossRef]
31. Abdel-Khalik, A.; Ahmed, S.; Massoud, A. Effect of multilayer windings with different stator winding connections on interior PM machines for EV applications. *IEEE Trans. Magn.* **2016**, *52*, 1–7. [CrossRef]
32. Ji, J.; Luo, J.; Zhao, W.; Zheng, J.; Zhang, Y. Effect of circumferential segmentation of permanent magnets on rotor loss in fractional-slot concentrated-winding machines. *IET Electr. Power Appl.* **2017**, *11*, 151–159. [CrossRef]
33. Cheng, M.; Zhu, S. Calculation of PM eddy current loss in IPM machine under PWM VSI supply with combined 2-D FE and analytical method. *IEEE Trans. Magn.* **2017**, *53*, 1–12. [CrossRef]
34. Xu, L.; Zhao, W.; Li, R.; Niu, S. Analysis of rotor losses in permanent magnet vernier machines. *IEEE Trans. Ind. Electron.* **2018**, *69*, 1224–1234. [CrossRef]

energies

MDPI

Article

Magnetic Equivalent Circuit and Optimization Method of a Synchronous Reluctance Motor with Concentrated Windings

Gan Zhang [1,*], Jinxin Tao [1], Yifan Li [1], Wei Hua [1], Xiaohan Xu [2] and Zhihong Chen [3]

1 School of Electrical Engineering, Southeast University, Nanjing 210096, China; 17851950582@163.com (J.T.); 15850656599@163.com (Y.L.); huawei1978@seu.edu.cn (W.H.)
2 State Grid Jiangsu Electric Power Co., Ltd. Maintenance Branch Company, Nanjing 211102, China; xxh-hayley@foxmail.com
3 Beijing Institute of Precision Mechatronics and Controls, Beijing 100076, China; zhihongch@outlook.com
* Correspondence: zhanggan@seu.edu.cn; Tel.: +86-1570-518-1487

Abstract: In this paper, the 12-slot/4-pole (12/4) synchronous reluctance motor (SynRM) with concentrated windings is proposed for low-cost hybrid vehicles. The non-linear magnetic equivalent circuit (MEC) model of the 12/4 SynRM is built to obtain the main electromagnetic characteristics such as coil flux, inductances, torque, etc. The magnetic saturation is also counted in by the MEC. Results calculated by MEC are validated by 2D finite element analysis (FEA). Then, aiming at larger average torque, lower torque ripple and lower total harmonic distortion (THD) in phase voltage, the parameter optimization method of the SynRM is proposed based on the Taguchi method and the MEC model. The proposed Taguchi-MEC method enables a fast optimization with satisfactory accuracy. Finally, the motor prototype is manufactured, and experimental validations are carried out.

Keywords: magnetic circuit; optimization; synchronous reluctance; torque performances

Citation: Zhang, G.; Tao, J.; Li, Y.; Hua, W.; Xu, X.; Chen, Z. Magnetic Equivalent Circuit and Optimization Method of a Synchronous Reluctance Motor with Concentrated Windings. *Energies* **2022**, *15*, 1735. https://doi.org/10.3390/en15051735

Academic Editor: Mario Marchesoni

Received: 18 January 2022
Accepted: 23 February 2022
Published: 25 February 2022

Publisher's Note: MDPI stays neutral with regard to jurisdictional claims in published maps and institutional affiliations.

1. Introduction

Thanks to the simple structure and robust rotor without a brush and slip ring, the synchronous reluctance motor (SynRM) has become an attractive candidate [1–3]. Conventional SynRMs employ distributed windings to obtain essentially sinusoidal back electromotive force (back-EMF) per phase and a high winding factor [4–7]. However, due to the long end-part winding length, distributed windings result in increased copper loss, decreased efficiency and power density. On the other hand, concentrated windings are rarely employed for SynRMs because of the high harmonic distortion in armature windings and low winding factor. Spargo and other authors introduce fractional slot concentrated windings (FSCW) to SynRM, with a segment stator to improve torque density and efficiency [8–14], e.g., the 6-slot 4-pole (6/4) SynRel machine. The 6/4 machine exhibits a comparable performance of a permanent magnet (PM)-assisted SynRM with a high torque output capability and a low level of torque ripple [8,9]. The proposed 12/4 CW machine exhibits comparable torque performance, compared to the case of adopting distributed winding, when the total stack length is unchanged. The short winding end, PM-free and very low mutual inductances of the 12/4 CW machine will be appreciated by applications where a short stack length, high robustness and low cost are required, such as the integrated starter generator (ISG) in low-cost wild hybrid vehicles. The ISG is located between the engine and the gear box, and drives the output shaft directly. First of all, the vehicle producer requires a brushless and PM-less motor and concentrated windings. Getting rid of permanent magnets reduces the cost and enhances robustness and speed range. Avoiding distributed windings improves the air flow and makes it more dustproof, as well as other mechanical considerations. These above requirements leave only the choices of the reluctance motor and SynRM. The 9/6 or 15/8 SynRM are not considered due to unbalanced radial forces on the rotor. More importantly, low mutual inductances between armature phases is highly appreciated, since it will facilitate a high fault-tolerant capability, which can be found in the 12/4 SynRM machine.

Consequently, the magnetic equivalent circuit (MEC) model is built to predict electromagnetic performance, including air-gap flux density distributions, winding inductances and electromagnetic torque. Then, the obtained results are validated by 2D-finite element analysis (2D-FEA). Furthermore, aiming at better torque performances and lower total harmonic distortion (THD) in phase voltage, the optimization of the 12/4 SynRM is carried out combing the MEC and Taguchi method. Unlike traditional optimization methods using FEA coupling optimization algorithms, such as a genetic algorithm [15–18] and particle swarm algorithm [19–22], the Taguchi method exhibits the merit of being less time consuming [23–26]. This merit can be greatly enhanced by using MEC instead of FEA, maintaining satisfactory accuracy.

2. Machine Topologies and Design Specifications

This section provides a brief introduction to a prototype of the 12/4 SynRM. The topology structure of the designed 12/4 SynRM is shown in Figure 1, with the key dimensional parameters listed in Table 1. The outer dimensions of the stator are chosen based on the available test bench and power converters for the following experiment. The segmented stator core is designed to improve the slot filling factor (SFF). Torque ripple reduction also needs to be mitigated, so that a continuous skew rotor is adopted. It should be noted that the skewing angle is optimized to achieve a balance between the sinusoidal back-EMF and electromagnetic torque. Two rotors are manufactured, one without skewing and another with skewing. This paper only introduces the 12/4 SynRM without rotor skewing. The skewed rotor will be introduced in another coming paper. Moreover, the seam/gap between the neighboring stator segments depends on the manufacturing accuracy and are normally less than 0.05 mm. Considering these seams are very small, their influences on magnetic performances can be neglected and not considered in the MEC and FEA models.

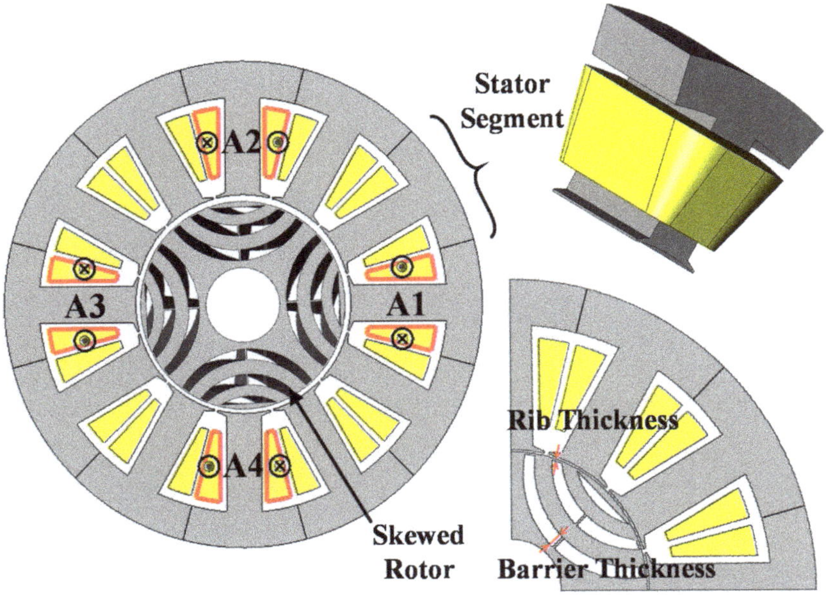

Figure 1. Topology of the 12/4 SynRM with concentrated windings with stator module and a quarter part.

Table 1. Main Design Specifications of the 12/4 Machine.

Item	Value
Outer diameter	150 mm
Lamination length	75 mm
Total end winding	20 mm
Total stack length	95 mm
β_2	12.4°
w_{b1}	10.00 mm
s_{tooth}	11.12 mm
s_{yoke}	11.38 mm
β_1	11.10°
d_{m1}	4.41 mm
$w_{rib1\text{-}1}$	0.58 mm
w_{rib2}	0.28 mm
d_{m2}	3.50 mm
$w_{rib1\text{-}2}$	0.50 mm
l_{so}	1.0 mm
Turns per coil	30
Copper fill factor	0.58
d-axis inductance (FEA)	19.3 mH
q-axis inductance (FEA)	6.2 mH
Phase current	30.1 A
Current angle	55°

3. Magnetic Equivalent Circuit Method and Evaluation

As an efficient tool, the MEC method has been widely utilized to analyze the electromagnetic performance of electrical machines [27,28], and is considered as a compromise between FEA with high accuracy and an analytical method (AM) with straightforward physical connections of performance and geometrical parameters. Therefore, the MEC model of the 12/4 SynRM is built in this paper. Figure 2 illustrates a quarter of the 12/4 machine, where each specific magnetic path is illustrated by a different MEC branch. The SynRM is separated into segments according to the geometrical shape. The equivalent permeance of each part of the stator and rotor core can be calculated using the permeance equations.

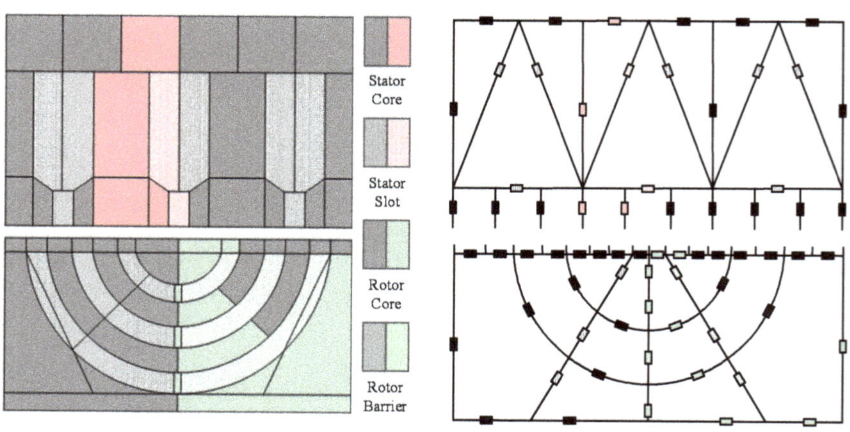

(**a**) (**b**)

Figure 2. MEC model for a quarter of 12/4 SynRM. (**a**) 1/4 topology, (**b**) 1/4 MEC model.

The basic equation which governs each branch of the magnetic circuit is given by,

$$\frac{\Phi}{G} = F \tag{1}$$

$$G = \mu_i \mu_0 \frac{S}{l} \tag{2}$$

In Equation (1), Φ, G and F are the flux, permeance and magnetic-motive-force (MMF), respectively. In Equation (2), μ_r, μ_0, S and l are the relative permeability, the permeability of free space, the cross-section area and the length, respectively. Specifically, μ_r is determined by iteration from the *B–H* curve of the lamination material. The permanent magnets can simply be modeled as an equivalent MMF [29],

$$F_m = \frac{B_r}{\mu_i \mu_0} h_m \tag{3}$$

and permeance,

$$G_m = \mu_i \mu_0 \frac{l_m l_a}{h_m} \tag{4}$$

where h_m and l_m are the magnet thickness and width, respectively, and B_r is the remanence.

Then, the MEC model is solved by MATLAB following the flow chart shown in Figure 3. The saturation in iron is counted in by the iteration of μ_r. For each branch in the magnetic circuit at the ith step in an iteration, the condition of convergence ε is defined as,

$$\varepsilon = \left| \frac{\mu_i - \mu_i'}{\mu_i} \right| \leq 10^{-6} \tag{5}$$

where μ_i is the permeance given before equation solving, and μ'_i is the real permeance obtained after the equation solving, as shown in Figure 4. If $\varepsilon \leq 10^{-6}$ is not achieved, the permeance at the $(i + 1)$th step in an iteration is given by Equation (6) and $k = 0.5$ in this case.

$$\mu_{i+1} = (1 - k)\mu_i + k\mu_i', \ 0 < k < 1 \tag{6}$$

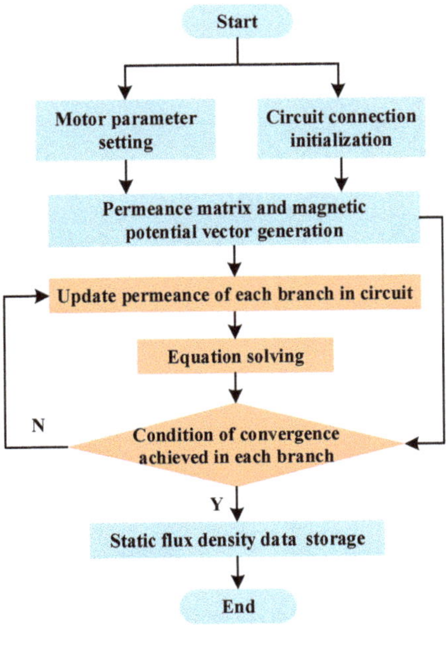

Figure 3. Calculation flow chart of the MEC model at each rotor position.

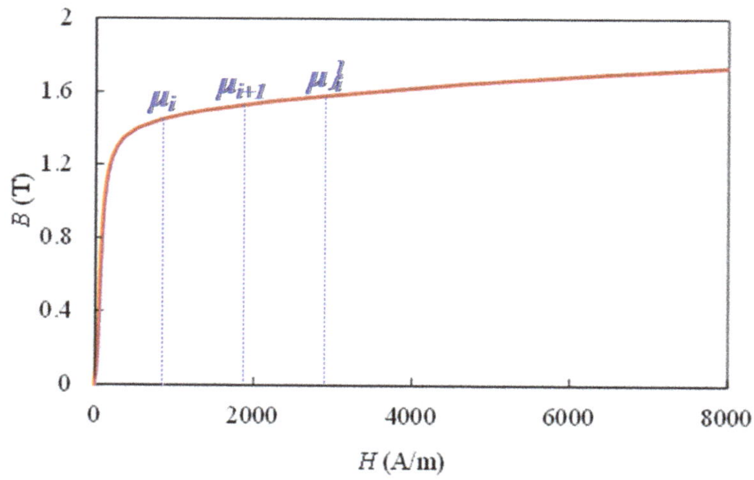

Figure 4. Demonstration of the convergence process according to the B–H curve.

Figure 5 shows the mesh results of the FEA model in which the air-gap consists of 6 layers of elements. The path to plot the air-gap flux density distribution is also given in Figure 5a, which covers 1/4 of the air-gap. The air-gap flux density distributions generated by phase-A current, d-axis current and q-axis current, respectively, are predicted by MEC and then compared to the FEA results by Figure 6. Overall, satisfied agreements are achieved.

(**a**) (**b**)

Figure 5. Elements in the FEA mesh. (**a**) The path to plot air-gap flux density, (**b**) mesh details.

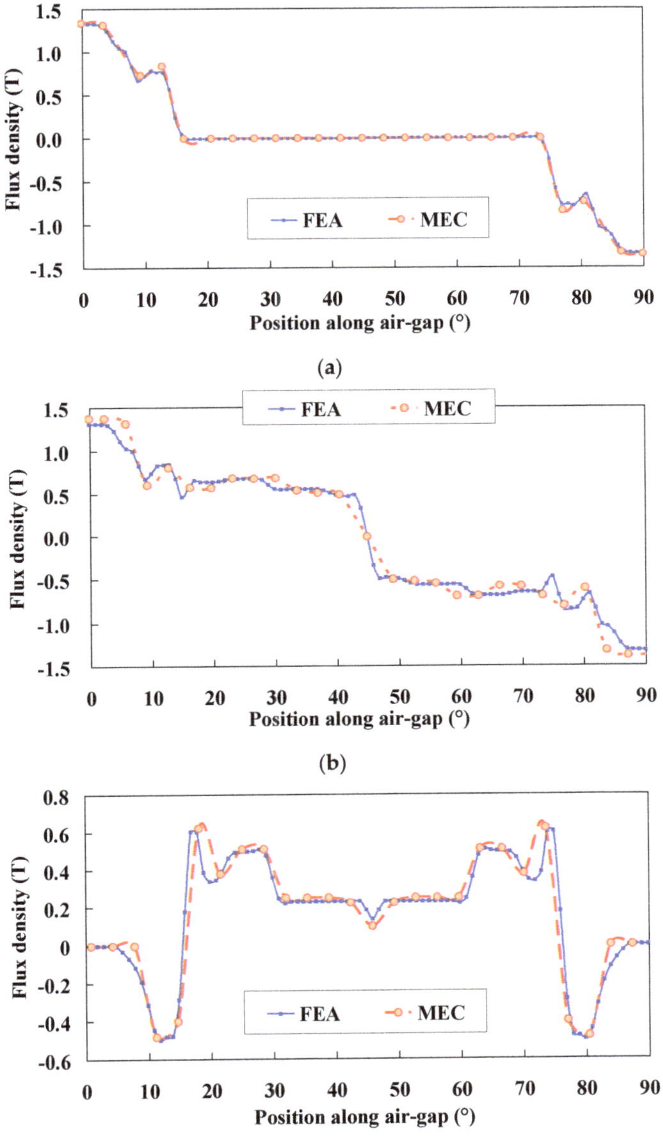

Figure 6. Air-gap flux density distributions: (**a**) by phase-A current; (**b**) by d-axis current; (**c**) by q-axis current.

4. Results and Discussion

When the MEC model of the 12/4 SynRM motor is solved, the electromagnetic characteristics can be easily calculated, e.g., the flux in the air-gap (Φ_A), coil flux linkage (Ψ_a), phase-A self-inductance (L_{aa}) and mutual inductance between phase-A and -B (M_{ba}) can be obtained by

$$\Phi_A = \int BdS \tag{7}$$

$$\Psi_A = N(\Phi_A + \Phi_{\delta A}) \tag{8}$$

$$L_{aa} = \frac{\Psi_A}{I_A} \tag{9}$$

$$M_{ba} = \frac{\Psi_B}{I_A} \tag{10}$$

Then, the L_d (d-axis), L_q (q-axis) and L_{dq} (d-q coupled inductance) can be calculated by Equations (11) and (12). Figure 7 compares the MEC and FEA obtained inductances with good agreements.

$$\begin{bmatrix} L_d \\ L_q \\ L_{dq} \end{bmatrix} = P \begin{bmatrix} L_{aa} & M_{ba} & M_{ca} \\ M_{ab} & L_{bb} & M_{cb} \\ M_{ac} & M_{bc} & L_{cc} \end{bmatrix} P^{-1} \tag{11}$$

$$P = \frac{2}{3} \begin{bmatrix} \cos\theta & \cos\left(\theta - \frac{2}{3}\pi\right) & \cos\left(\theta + \frac{2}{3}\pi\right) \\ -\sin\theta & -\sin\left(\theta - \frac{2}{3}\pi\right) & -\sin\left(\theta + \frac{2}{3}\pi\right) \\ \frac{1}{2} & \frac{1}{2} & \frac{1}{2} \end{bmatrix} \tag{12}$$

$$P^{-1} = \begin{bmatrix} \cos\theta & -\sin\theta & \frac{1}{2} \\ \cos\left(\theta - \frac{2}{3}\pi\right) & -\sin\left(\theta - \frac{2}{3}\pi\right) & \frac{1}{2} \\ \cos\left(\theta + \frac{2}{3}\pi\right) & -\sin\left(\theta + \frac{2}{3}\pi\right) & \frac{1}{2} \end{bmatrix} \tag{13}$$

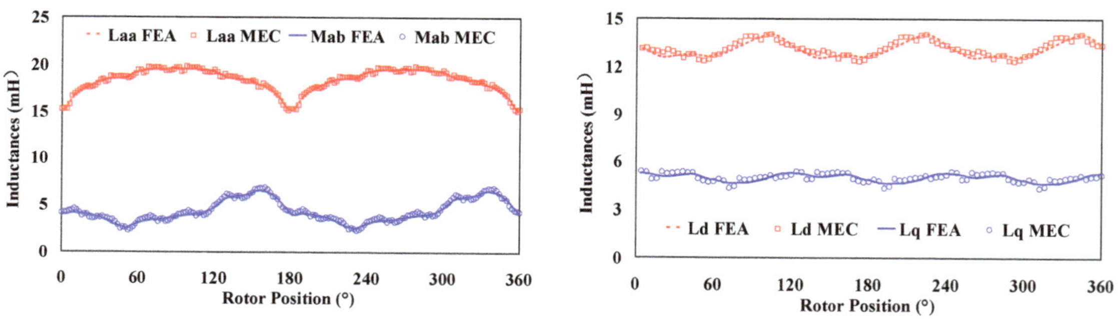

(a) (b)

Figure 7. Comparison of the inductance results obtained by MEC and FEA. (**a**) Phase self and mutual inductances when 1 A is applied in phase-A. (**b**) d-axis and q-axis inductances at rated operation.

It should be noted that in the 12/4 SynRM with concentrated windings, there is a cross-coupling between the d-axis and q-axis, i.e., the L_{dq} contributes unneglectable electromagnetic torque, as given in Equation (14).

$$T = \frac{3}{2} p {i_s}^2 \left[\frac{1}{2} \left(L_d - L_q \right) \sin 2\alpha + L_{dq} \cos 2\alpha \right] \tag{14}$$

where p is the rotor pole-pairs, L_d and L_q is the d- and q-axis inductance, respectively, and α is the current phase angle, i.e., the angle between the current vector and d-axis. Thus, the torque waveform can be calculated. Figures 8 and 9 give the torque performance versus current phases and rotor position, respectively. As can be seen, good agreements are achieved between the MEC and FEA results.

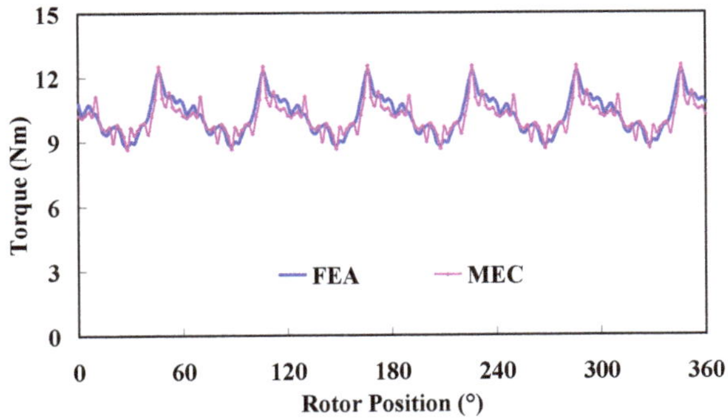

Figure 8. Torque vs. rotor positions when 30 A is applied.

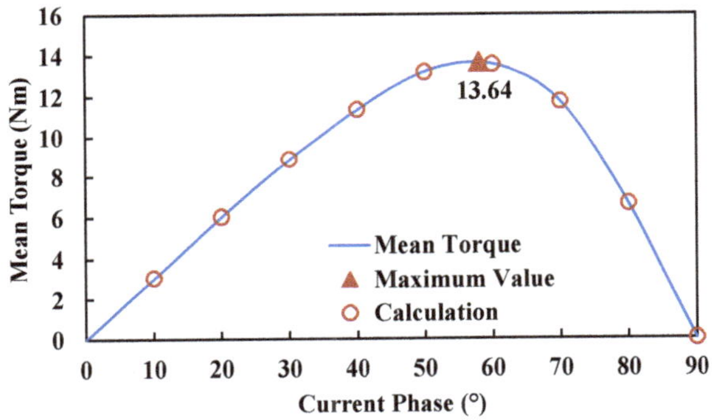

Figure 9. Average torque, calculation values vs. current phases when 50 A is applied.

5. Optimization of Torque Characters Based on MEC and Taguchi Method

The Taguchi method is chosen as one of the many optimization methods. The optimization methods introduced in Section 1 require sophisticated algorithms and additional programming and data processing along with a lot of FEA cases of electromagnetic field. So, the methods are usually time consuming and complex. On the other hand, the Taguchi method can obtain satisfactory results by reduced calculation cases, and thus is less time consuming. In this paper, the Taguchi method and the MEC model are used to optimize the SynRM. The steps in the Taguchi method are as follows [30]:

1. Select the parameters and determine its levels.
2. Design of exams.
3. Review the optimization results.

First of all, the means of all results $M_t(S)$ and average effect M_{X_j} can be calculated as

$$M_t(S) = \frac{1}{n} \sum_{p_n=1}^{n} S_{p_n} \tag{15}$$

$$M_{X_j}(Y) = \frac{1}{3}(Y_1 + Y_2 + Y_3) \tag{16}$$

where X_i and Y_i are factor-level and performances, respectively. Then, the analysis of variance is carried out, which provides a measure of confidence. The technique does not directly analyze the data, but rather determines the variance of the data. The sum of squares

(SS) is calculated first. It is a measure of the deviation of the experimental data from the mean value of the data. The sum of squares due to various factors can be calculated as

$$SS = j\sum_{p_j=1}^{j}\left(M_{X_{p_j}}(Y) - M_t(S)\right)^2 \tag{17}$$

where $M_t(S)$ and $M_{xpj}(Y)$ are results of Equations (5) and (6). SS shows the relative importance of various factors on machine performance. Then, the Pearson coefficient is chosen to calculate the coefficient of correlation and given as

$$\rho_{X_i,Y_i} = \frac{n\sum X_iY_i - \sum X_i \sum Y_i}{\sqrt{n\sum X_i^2 - (\sum X_i)^2}\sqrt{n\sum Y_i^2 - (\sum Y_i)^2} } \tag{18}$$

The optimization aim is a larger average torque (T_{avg}), lower torque ripple (T_{rip}) and lower THD of phase voltage (THD_U). Firstly, the optimization parameters are determined and shown in Figure 10 and listed in Table 2, which are related to the rotor barriers and ribs, stator tooth and yoke width.

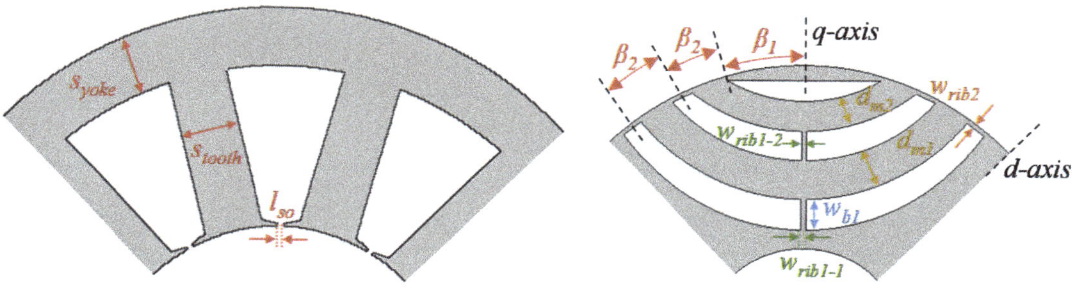

(a) (b)

Figure 10. Design parameters to be optimized. (**a**) Stator part. (**b**) Rotor part.

Table 2. Design Parameters and Their Variation Levels.

Parameter	Level 1	Level 2	Level 3
β_1	11.5°	12.5°	12°
β_2	11.5°	12.5°	12°
d_{m1}	3.5 mm	4.5 mm	4 mm
d_{m2}	2.5 mm	3.5 mm	3 mm
w_{b1}	3 mm	4 mm	3.5 mm
w_{rib1}	0.55 mm	0.65 mm	0.6 mm
w_{rib2}	0.45 mm	0.55 mm	0.5 mm
s_{yoke}	10 mm	12 mm	11 mm
s_{tooth}	10 mm	12 mm	11 mm
l_{so}	0.8 mm	1.2 mm	1 mm

The design of the exam is carried out after preparing the initial information. To obtain the number of exams, the orthogonal arrays are presented using Taguchi. The exams' matrix can be easily created using existing software, in this case the Minitab. Shown in Table 3 are the Taguchi experiments that are tailored to their parameters and levels. The rows of the exams table represent the levels of factors in each experiment, and its columns represent the number of factors.

Table 3. Exams Table.

Exams	A	B	C	D	E	F	G	H	J	K
1	1	1	1	1	1	1	1	1	1	1
2	1	1	1	1	2	2	2	2	2	2
3	1	1	1	1	3	3	3	3	3	3
4	1	2	2	2	1	1	1	2	2	2
5	1	2	2	2	2	2	2	3	3	3
6	1	2	2	2	3	3	3	1	1	1
7	1	3	3	3	1	1	1	3	3	3
8	1	3	3	3	2	2	2	1	1	1
9	1	3	3	3	3	3	3	2	2	2
10	2	1	2	3	1	2	3	1	2	3
11	2	1	2	3	2	3	1	2	3	1
12	2	1	2	3	3	1	2	3	1	2
13	2	2	3	1	1	2	3	2	3	1
14	2	2	3	1	2	3	1	3	1	2
15	2	2	3	1	3	1	2	1	2	3
16	2	3	1	2	1	2	3	3	1	2
17	2	3	1	2	2	3	1	1	2	3
18	2	3	1	2	3	1	2	2	3	1
19	3	1	3	2	1	3	2	1	3	2
20	3	1	3	2	2	1	3	2	1	3
21	3	1	3	2	3	2	1	3	2	1
22	3	2	1	3	1	3	2	2	1	3
23	3	2	1	3	2	1	3	3	2	1
24	3	2	1	3	3	2	1	1	3	2
25	3	3	2	1	1	3	2	3	2	1
26	3	3	2	1	2	1	3	1	3	2
27	3	3	2	1	3	2	1	2	1	3

Using the results of experiments designed using the Taguchi method and after analyzing the results of the experiments, the optimal combination of factor levels and output voltage and cost values are calculated at the optimal point. The effect of each level is listed in Table 4.

Table 4. The Average Value of Levels.

Value	T_{avg} (Nm)			T_{rip}			THD_U		
Level	1	2	3	1	2	3	1	2	3
β_1	9.80	9.82	9.79	0.43	0.41	0.41	0.41	0.37	0.39
β_2	9.57	9.97	9.82	0.55	0.32	0.35	0.39	0.38	0.39
d_{m1}	10.12	9.80	9.77	0.34	0.44	0.44	0.39	0.38	0.39
d_{m2}	9.72	9.68	9.85	0.41	0.43	0.43	0.39	0.38	0.39
w_{b1}	9.26	10.13	9.61	0.42	0.43	0.44	0.39	0.37	0.39
w_{rib1}	9.64	9.99	9.73	0.38	0.41	0.39	0.35	0.42	0.36
w_{rib2}	9.93	9.85	9.81	0.43	0.43	0.42	0.38	0.39	0.38
s_{yoke}	9.98	9.42	9.86	0.42	0.42	0.41	0.39	0.38	0.39
s_{tooth}	9.58	9.99	9.87	0.43	0.41	0.44	0.41	0.36	0.38
l_{so}	9.82	9.88	9.83	0.41	0.45	0.43	0.38	0.38	0.38

Figures 11–13 present the output graph for different levels of each parameter. It can be observed that T_{avg} is most sensitive to S_{yoke} and S_{tooth}, while T_{rip} is most sensitive to β_2, w_{b1} and S_{tooth}. On the other hand, w_{b1} and S_{tooth} have the greatest influence on THD_U.

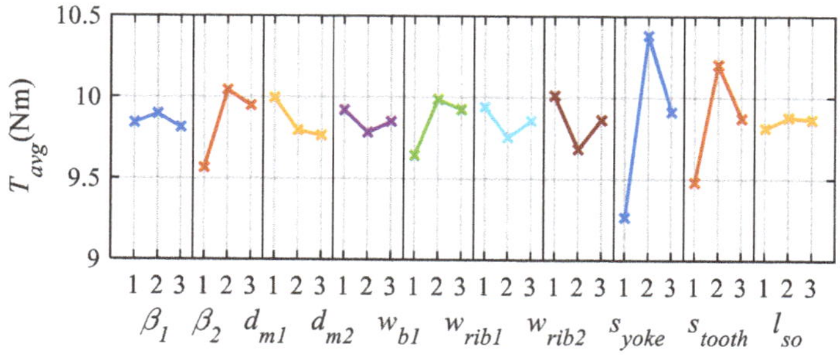

Figure 11. Plot of main factor effects on average torque (T_{avg}).

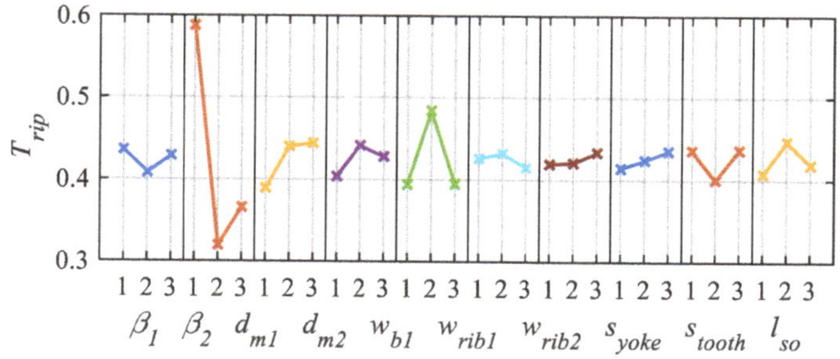

Figure 12. Plot of main factor effects on torque ripple (T_{rip}).

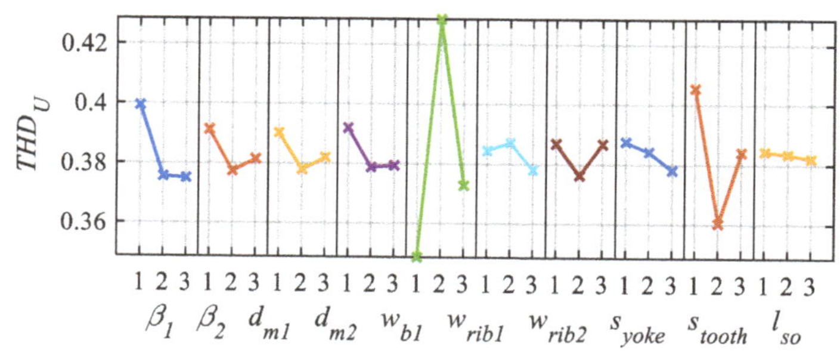

Figure 13. Plot of main factor effects on THD of phase voltage (THD_U).

Table 5 gives the initial parameters and Taguchi-MEC optimized parameters. Table 6 compares the T_{avg}, T_{rip} and THD_U. As can be seen, the final results obtained by Taguchi-MEC are very close to those obtained by the multi-object optimization carried out on JMAG. It should be emphasized that the total time needed for Taguchi-MEC is 10 min (including nearly 8 min organizing data), while the time needed for JMAG is nearly 58 h (43,086 elements in mesh and 180 steps in each case). Overall, the proposed Taguchi-MEC method exhibits the great advantage of fast calculation with promising accuracy.

A brief comparison of the proposed 12/4 SynRM with existing state of the art is carried out. Figure 14 shows the 12/4 SynRM with concentrated winding (CW) and its counterpart, i.e., the 12/4 SynRM with distributed windings (DW). As can be seen, the end winding in the CW machine is much smaller than the DW one; thus, more laminations can be adopted when the same stack length is achieved. A higher copper fill factor and more coil turns are also obtained by the CW machine. More details regarding design specification and electromagnetic performances are given in Table 7, where the DW SynRM is optimized by a multi-object optimization method carried out on JMAG. The fault tolerant capability is evaluated by the mutual linkage coefficient k_{ml}, which is also given in Table 7.

Table 5. Comparison of the Parameters and Performances.

Parameters	Initial	Optimized by Taguchi-MEC
β_2	12.0°	12.4°
w_{b1}	3.50 mm	10.00 mm
s_{tooth}	11.00 mm	11.12 mm
s_{yoke}	11.00 mm	11.38 mm
β_1	12.00°	11.10°
d_{m1}	3.50 mm	4.41 mm
$w_{rib1\text{-}1}$	0.60 mm	0.58 mm
w_{rib2}	0.50 mm	0.28 mm
d_{m2}	2.50 mm	3.50 mm
$w_{rib1\text{-}2}$	0.60 mm	0.50 mm
l_{so}	1.0 mm	1.0 mm

Table 6. Comparison of the Optimization Results Between Taguchi-MEC and FEA.

Results	Initial	Taguchi-MEC	FEA (by JMAG)
T_{avg}	9.40 Nm	9.82 Nm	9.90 Nm
T_{rip}	37.4%	31.2%	30.5%
THD_U	37%	35%	35%

$$k_{ml} = \frac{\sum_{\theta=0}^{360}(|M_{ab}|)}{\sum_{\theta=0}^{360}(L_{aa})} \tag{19}$$

where θ is electrical rotor position. The smaller k_{ml} indicates weaker mutual linkage between armature phases and thus better fault tolerant capability. As can be seen from Table 7, the proposed 12/4 CW SynRM shows a better torque output result and much better k_{ml} than the traditional 12/4 DW SynRM. The advantage of the torque performances of the CW machine is more obvious when the total stack length is smaller than 95 mm since the proportion of lamination length to total stack length varies with available space in stack.

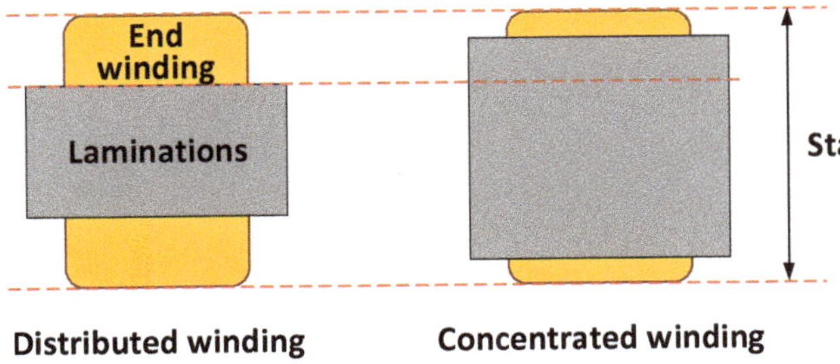

Figure 14. Demonstration of the end winding of the 12/4 SynRMs.

Table 7. Design Specifications of the 12/4 SynRMs.

Item	Distributed Winding	Concentrated Winding
Outer diameter	150 mm	150 mm
Lamination length	50 mm	75 mm
Total end winding	45 mm	20 mm
Total stack length	95 mm	95 mm
Turns per coil	30	40
Copper fill factor	0.5	0.58
Average torque	9.6 Nm	9.9 Nm
Torque ripple	30%	30.5%
Mutual linkage coefficient, k_{ml}	50%	14%

6. Experimental Validation

The experimental setup shown in Figure 15 consists of a prototyped 12/4 SynRM and a back-to-back load machine, which are both controlled by inverters. The load machine of the experimental platform supports the torque range of 50 Nm and the max speed of 15,000 r/min. The inverter supports the power of 30 kW/60 A. The torque sensor supports the torque range of 50 Nm/10 Nm with the accuracy of 0.1%, respectively. The torque shown in Figure 16 is measured by a sensor positioned between the SynRM and the load machine. Figure 17 compares the torque values obtained by MEC and FEA, and as can be seen, good agreements are achieved. The FEA calculated losses and efficiency is given in Table 8, and the measured efficiency under different rotor speed is presented in Figure 18.

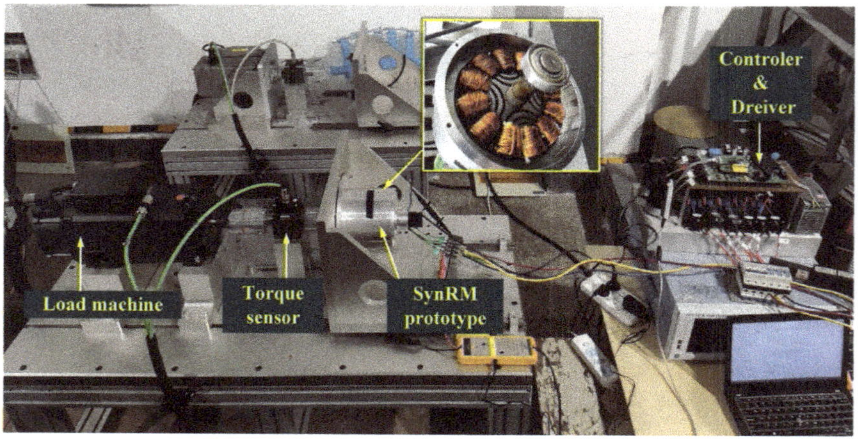

Figure 15. Experimental platform.

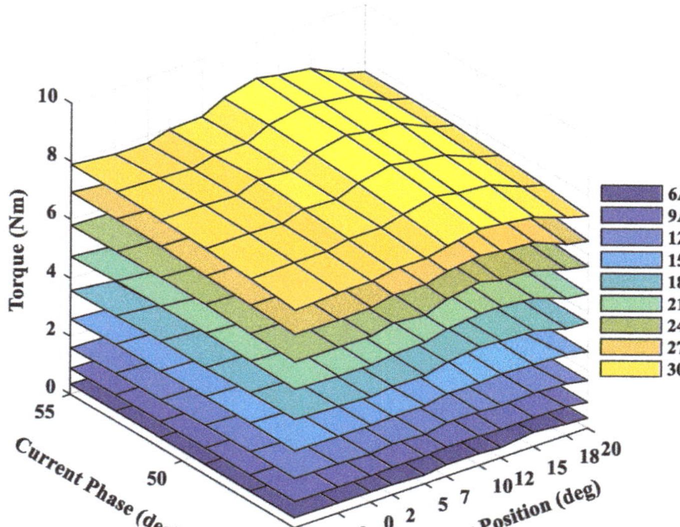

Figure 16. Measured torque vs. current phases and rotor positions.

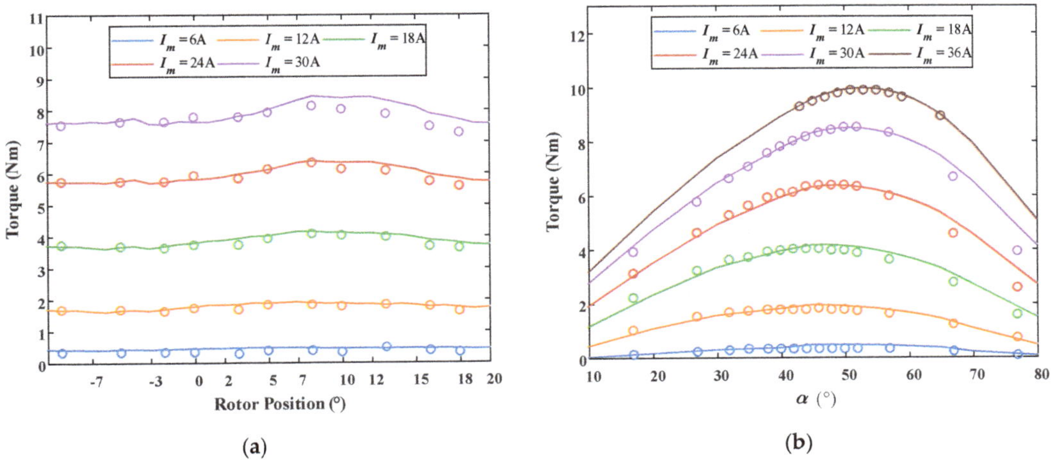

Figure 17. Comparison of the measured and FEA calculated torque results: (**a**) torque values by different current and rotor position; (**b**) torque values by different current and current angle.

Table 8. Power loss and efficiency at rated operation.

Item	FEA Results
Iron loss	15.7 W
Copper loss	130.8 W
Efficiency	91.40%
Power factor	0.65
Speed	1500 r/min

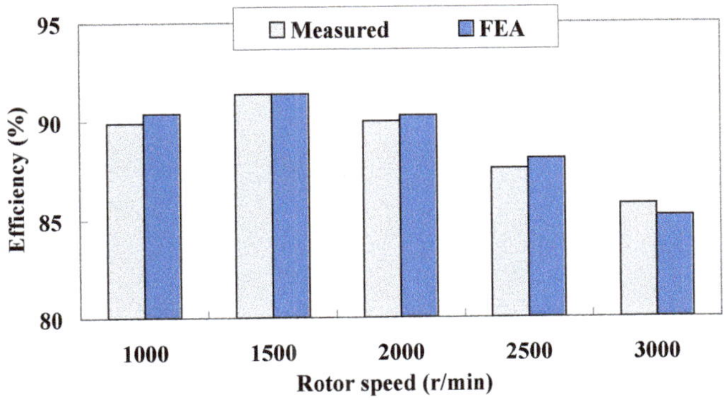

Figure 18. Comparison of the measured and FEA calculated efficiency by different rotor speed.

7. Conclusions

This paper demonstrates the electromagnetic performance of a 12/4 SynRM with concentrated windings. The non-linear MEC model of the 12/4 SynRM is built to get the air-gap flux density, coil flux-linkages, armature winding inductances, etc. The saturation effect is counted in by the MEC. The MEC model is validated by 2D-FEA. Then, the parameter optimization method based on Taguchi and MEC is proposed, aiming at the best average torque, lowest torque ripple and THD in phase voltage. Overall, the proposed Taguchi-MEC method exhibits comparable accuracy with traditional FEA-based multi-object parameter optimization methods, but using much less time. It should be emphasized that the MEC is not used to deal with complex topologies such as a rotor with a non-uniform rotor iron shape or rotor outline. Nevertheless, the proposed MEC-Taguchi method can be employed to obtain a roughly optimized topology that can be used as the initial design to be further optimized by FEA.

Author Contributions: Conceptualization, G.Z.; data curation, Y.L., G.Z. and J.T.; formal analysis, G.Z. and Z.C.; funding acquisition, Z.C. and W.H.; investigation, G.Z. and Y.L.; methodology, G.Z., X.X. and Y.L.; supervision, G.Z. and Z.C.; writing—original draft, G.Z. All authors have read and agreed to the published version of the manuscript.

Funding: This work is supported in part by the National Nature Science Foundation of China under Grant 51937006, 52077032, 51991380, in part by CALT under Grant CALT201806, in part by the Fundamental Research Funds for the Central Universities under Grant 2242021R41129, and in part by the Technology Foundation for Selected Overseas Chinese Scholar under Grant 1116000256.

Conflicts of Interest: The authors declare no conflict of interest.

References

1. Bianchi, N.; Bolognani, S.; Bon, D.; Pré, M.D. Torque harmonic compensation in a synchronous reluctance motor. *IEEE Trans. Energy Convers.* **2008**, *23*, 466–473. [CrossRef]
2. Bianchi, N.; Bolognani, S.; Bon, D.; Pre, M.D. Rotor flux-barrier design for torque ripple reduction in synchronous reluctance and PM-assisted synchronous reluctance motors. *IEEE Trans. Ind. Appl.* **2009**, *45*, 921–928. [CrossRef]
3. Lopez-Torres, C.; Bacco, G.; Bianchi, N.; Espinosa, A.G.; Romeral, L. A parallel analytical computation of synchronous reluctance machine. In Proceedings of the 2018 XIII International Conference on Electrical Machines (ICEM), Alexandroupoli, Greece, 3–6 September 2018; pp. 25–31.
4. Babetto, C.; Bacco, G.; Bianchi, N. Synchronous reluctance machine optimization for high-speed applications. *IEEE Trans. Energy Convers.* **2018**, *33*, 1266–1273. [CrossRef]
5. Murataliyev, M.; Degano, M.; Galea, M. A novel sizing approach for synchronous reluctance machines. *IEEE Trans. Ind. Electron.* **2021**, *68*, 2083–2095. [CrossRef]
6. Moghaddam, R.; Gyllensten, F. Novel high-performance synrm design method: An easy approach for a complicated rotor topology. *IEEE Trans. Ind. Electron.* **2014**, *61*, 5058–5065. [CrossRef]
7. Spargo, S.M.; Mecrow, B.C.; Widmer, J.D.; Morton, C.; Baker, N.J. Design and validation of a synchronous reluctance motor with single tooth windings. *IEEE Trans. Energy Convers.* **2015**, *30*, 795–805. [CrossRef]
8. Spargo, C.M.; Mecrow, B.C.; Widmer, J.D.; Morton, C. Application of fractional-slot concentrated windings to synchronous reluctance motors. *IEEE Trans. Ind. Appl.* **2015**, *51*, 1446–1455. [CrossRef]
9. Spargo, C.M.; Mecrow, B.C.; Widmer, J.D. Electromagnetic analysis of a synchronous reluctance motor with single-tooth windings. *IEEE Trans. Magn.* **2017**, *53*, 8206207.
10. Spargo, C.M. Electromagnetic–mechanical design of synchronous reluctance rotors with fine features. *IEEE Trans. Magn.* **2017**, *53*, 8206308.
11. Babetto, C.; Bianchi, N.; Torreggiani, A.; Bianchini, C.; Davoli, M.; Bellini, A. Optimal design and experimental validation of a synchronous reluctance machine for fault-tolerant applications. In Proceedings of the 2019 IEEE Energy Conversion Congress and Exposition (ECCE), Baltimore, MD, USA, 29 September–3 October 2019; pp. 4880–4887.
12. Lehner, B.; Gerling, D. Design considerations for concentrated winding synchronous reluctance machines. In Proceedings of the 2016 IEEE Transportation Electrification Conference and Expo, Asia-Pacific (ITEC), Busan, Korea, 1–4 June 2016; pp. 485–490.
13. Gamba, M.; Pellegrino, G.; Armando, E.; Ferrari, S. Synchronous reluctance motor with concentrated windings for IE4 efficiency. In Proceedings of the 2017 IEEE Energy Conversion Congress and Exposition (ECCE), Cincinnati, OH, USA, 1–5 October 2017; p. 17351585.
14. Ma, X.Y.; Li, G.J.; Zhu, Z.Q.; Jewell, G.W.; Green, J. Investigation on synchronous reluctance machines with different rotor topologies and winding configurations. *IET Electr. Power Appl.* **2018**, *12*, 45–53. [CrossRef]
15. Jannot, X.; Vannier, J.; Marchand, C.; Gabsi, M.; Michel, J.S.; Sadarnac, D. Multiphysic modeling of a high-speed interior permanentmagnet synchronous machine for a multiobjective optimal design. *IEEE Trans. Energy Convers.* **2011**, *26*, 457–467. [CrossRef]
16. Nakata, T.; Sanada, M.; Morimoto, S.; Inoue, Y. Automatic design of IPMSMs using a genetic algorithm combined with the coarse-mesh FEM for enlarging the high-efficiency operation area. *IEEE Trans. Ind. Electron.* **2017**, *64*, 9721–9728. [CrossRef]
17. Rostami, N.; Feyzi, M.R.; Pyrhonen, J.; Parviainen, A.; Behjat, V. Genetic algorithm approach for improved design of a variable speed axial-flux permanent-magnet synchronous generator. *IEEE Trans. Magn.* **2012**, *48*, 4860–4865. [CrossRef]
18. Ismail, M.M.; Xu, W.; Wang, X.G.; Junejo, A.K.; Liu, Y.; Dong, M.H. Analysis and optimization of torque ripple reduction strategy of surface-mounted permanent-magnet motors in flux-weakening region based on genetic algorithm. *IEEE Trans. Ind. Appl.* **2021**, *57*, 4091–4106. [CrossRef]
19. Dong, F.; Zhao, J.; Song, J.; Feng, Y.; He, Z. Optimal design of permanent magnet linear synchronous motors at multispeed based on particle swarm optimization combined with SN ratio method. *IEEE Trans. Energy Convers.* **2018**, *33*, 1943–1954. [CrossRef]

20. Lee, J.H.; Song, J.; Kim, D.; Kim, J.; Kim, Y.; Jung, S. Particle swarm optimization algorithm with intelligent particle number control for optimal design of electric machines. *IEEE Trans. Ind. Electron.* **2018**, *65*, 1791–1798. [CrossRef]
21. Lin, F.; Teng, L.; Chu, H. A robust recurrent wavelet neural network controller with improved particle swarm optimization for linear synchronous motor drive. *IEEE Trans. Power Electron.* **2008**, *23*, 3067–3078. [CrossRef]
22. Chen, Z.; Li, W.C.; Shu, X.; Shen, J.W.; Zhang, Y.J.; Shen, S.Q. Operation efficiency optimization for permanent magnet synchronous motor based on improved particle swarm optimization. *IEEE Access* **2020**, *9*, 777–788. [CrossRef]
23. Lee, S.J.; Kim, K.S.; Cho, S.; Jang, J.; Lee, T.H.; Hong, J.P. Optimal design of the magnetizing fixture using Taguchi robust design in the ringtype PMSM. In Proceedings of the 7th IET International Conference on Power Electronics, Machines and Drives (PEMD 2014), Manchester, UK, 8–10 April 2014; pp. 1–6.
24. Dong, F.; Song, J.; Zhao, J.; Zhao, J. Multi-objective design optimization for PMSLM by FITM. *IET Electric Power Appl.* **2018**, *12*, 188–194. [CrossRef]
25. Si, J.; Zhao, S.; Feng, H.; Cao, R.; Hu, Y. Multi-objective optimization of surface-mounted and interior permanent magnet synchronous motor based on Taguchi method and response surface method. *Chin. J. Elect. Eng.* **2018**, *4*, 67–73.
26. Shi, Z.; Sun, X.D.; Cai, Y.F.; Yang, Z.B. Robust design optimization of a five-phase pm hub motor for fault-tolerant operation based on Taguchi method. *IEEE Trans. Energy Convers.* **2020**, *35*, 2036–2044. [CrossRef]
27. Amrhein, M.; Krein, P.T. Magnetic equivalent circuit simulations of electrical machines for design purposes. In Proceedings of the 2007 IEEE Electric Ship Technologies Symposium, Arlington, VA, USA, 21–23 May 2007; pp. 254–260.
28. Cao, D.; Zhao, W.; Ji, J.; Ding, L.; Zheng, J. A generalized equivalent magnetic network modeling method for vehicular dual-permanent-magnet vernier machines. *IEEE Trans. Energy Convers.* **2019**, *34*, 1950–1962. [CrossRef]
29. Zhu, Z.Q.; Pang, Y.; Howe, D.; Iwasaki, S.; Deodhar, R.; Pride, A. Analysis of electromagnetic performance of flux-switching permanent-magnet machines by nonlinear adaptive lumped parameter magnetic circuit model. *IEEE Trans. Magn.* **2005**, *41*, 4277–4287. [CrossRef]
30. Alireza, S.; Mahsa, K. Optimum design of double sided linear switched reluctance motor with Taguchi method. In Proceedings of the 2019 IEEE 28th International Symposium on Industrial Electronics (ISIE) Vancouver, Vancouver, BC, Canada, 12–14 June 2019; p. 18870309.

energies

Article

Parameter Sensitivity Analysis and Robust Design Approach for Flux-Switching Permanent Magnet Machines

Gan Zhang [1,*], Qing Tong [1], Anjian Qiu [1], Xiaohan Xu [2], Wei Hua [1] and Zhihong Chen [3]

[1] School of Electrical Engineering, Southeast University, Nanjing 210096, China; qtong_ss@seu.edu.cn (Q.T.); 220213122@seu.edu.cn (A.Q.); huawei1978@seu.edu.cn (W.H.)
[2] Maintenance Branch Company, State Grid Jiangsu Electric Power Co., Ltd., Nanjing 211102, China; xuxh3@js.sgcc.com.cn
[3] Beijing Institute of Precision Mechatronics and Controls, Beijing 100076, China; chenzhh@casc.vip
* Correspondence: zhanggan@seu.edu.cn; Tel.: +86-157-0518-1487

Abstract: Parameter sensitivity analysis is usually required to select the key parameters with high sensitivity to the optimal goal before the optimization is carried out, especially for flux-switching permanent magnet (FSPM) machines where lot of design parameters should be considered. Unlike the traditional studies on parameter sensitivity, which are generally experience- or statistics-based, and are time-consuming, this paper proposes a parameter sensitivity analysis method of a FSPM machine based on a magnetic equivalent circuit (MEC), which enables the parameters' sensitivities to be evaluated by their exponential in the nondimensionalized equations, thus providing a fast and accurate way to obtain the parameter sensitivities. Thereafter, the influences of modular manufacturing methods on magnetic performances are discussed, and the robust design approach for the FSPM machine is introduced, which aims to achieve the best machine stability and robustness by setting boundaries on design dimensions when taking into account the manufacturing tolerances. Experimental validations are also presented.

Keywords: flux switching; parameter sensitivity; robust design approach; stator-PM

Citation: Zhang, G.; Tong, Q.; Qiu, A.; Xu, X.; Hua, W.; Chen, Z. Parameter Sensitivity Analysis and Robust Design Approach for Flux-Switching Permanent Magnet Machines. *Energies* **2022**, *15*, 2194. https://doi.org/10.3390/en15062194

Academic Editors: Massimo Panella and J. C. Hernandez

Received: 31 January 2022
Accepted: 15 March 2022
Published: 17 March 2022

1. Introduction

Flux-switching permanent magnet (FSPM) machines exhibit high torque and power densities [1–4], which endows them with great potential in applications such as electric vehicles (EV) and hybrid EVs [5,6]. Locating both the magnets and armature windings on the stator will facilitate the accommodation of a water-cooling system and modular manufacturing method. Lots of work regarding the design method have been carried out [7–9], as well as the calculating methods, e.g., the field modulation theory [10–12], the lumped parameter-based magnetic circuit model [13], as well as Fourier analysis-based methods [14,15]. The irreversible demagnetization of FSPM machine is also discussed in [16]. As for the optimization method, the multiobjective optimization method is adopted by the flux-switching machines [17–21], e.g., response surface analysis [18] and genetic algorithm optimization [19]. The sensitivity analysis is defined as a technique that determines how different values of an independent variable can impact a dependent variable under a given set of assumptions [22,23]. Usually, parameter sensitivity analysis is needed to select key parameters with high sensitivity to the optimal goal before the optimization is carried out. The sensitive value represents the correlation degree between the optimal goal and the dimension variable. Traditional studies on parameter sensitivity are experience- or statistics-based [19,24,25], where lots of finite-element analysis (FEA) calculation is required, making them time-consuming. In this paper, the parameter sensitivity of the FSPM machines is analyzed based on a magnetic equivalent circuit (MEC) method where FEA calculation for only a few cases is needed. The proposed MEC-based method provides a fast and accurate way to obtain the parameter sensitivity results. Specifically, based on a 12-stator-slot/10-rotor-pole (12/10) FSPM machine and 12/14 FSPM machine as shown in Figure 1a,b, the relationship between the electromagnetic performances and the key design parameters shown in Figure 1c and Table 1 are given by equations in the nondimensionalized form, which are deduced from the simplified equivalent magnetic circuit (SEMC) of the FSPM machine, and validated by FEA results. Thus, the parameters' sensitivities are quantitatively evaluated by their exponential in the nondimensionalized equations. A greater exponential means the design parameter has higher impact on the corresponding electromagnetic performances; thus, it is a dominant parameter and should be given more weight during optimization. It should be emphasized that, instead of very complicated equations, using nondimensionalized form enables the parameter sensitivities to be presented in a much easier and simplified way. Additionally, the modular manufacturing method of the FSPM machine is introduced to facilitate the fabricating process and obtain better copper fill result, as shown in Figure 2. However, this manufacturing method might generate larger tolerances than traditional methods; thus, its influences on electromagnetic performances are investigated. Figure 3 gives the overall flowchart of the parameter sensitivity analysis process. The robust design approach for the FSPM machine are discussed [26], which aims at best machine stability and robustness when taking into account the manufacturing tolerances. Finally, the experimental validations are carried out.

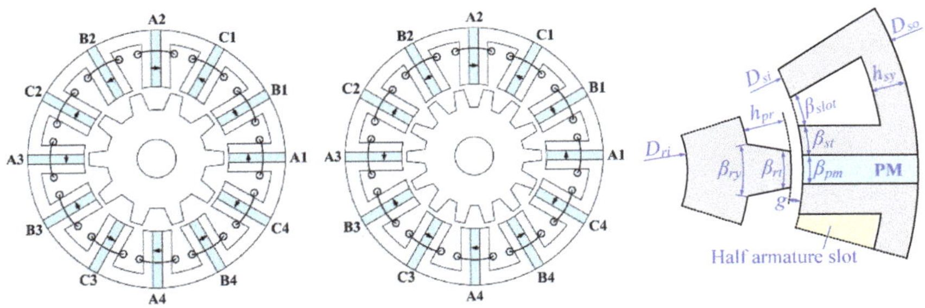

(a) (b) (c)

Figure 1. Configuration of the FSPM machines. (**a**) 12/10. (**b**) 12/14. (**c**) Design parameters.

Table 1. Main design specifications.

Symbol	Parameter	Quantity
n_N	Rated rotor speed	1000 r/min
B_r	PM remanence at 25 °C	1.2 T
g	Air-gap length	0.9 mm
P_s	Stator slot number	12
P_r	Rotor pole number	10
P_n	Rated power	4.8 kW
I_a	Rated current	60 Arms
T_n	Rated torque	46 Nm
N_{coil}	Turns per coil	18
D_{so}	Stator outer diameter	240 mm
D_{si}	Stator inner diameter	14 mm
k_{sio}	Stator split ratio	0.6
l_a	Active stack length	40 mm
β_{pm}	Magnet width arc	6.0°
β_{st}	Stator tooth arc	8.0°
β_{slot}	Stator slot arc	8.0°
h_{sy}	Stator yoke width	9.4 mm
β_{rt}	Rotor tooth arc	10.5°
β_{ry}	Rotor tooth yoke arc	21.0°
h_{pr}	Rotor yoke width	17.8 mm

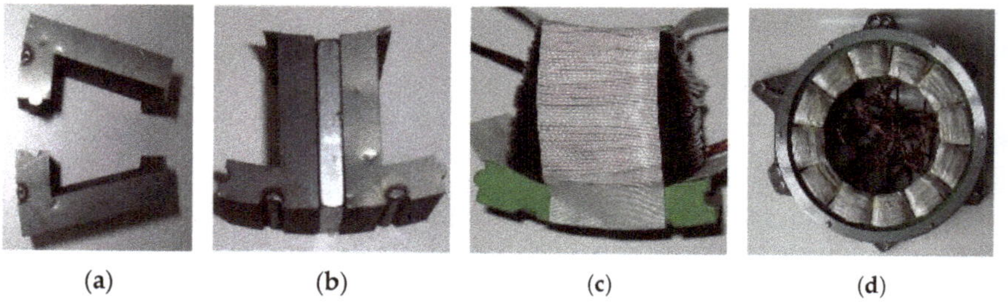

Figure 2. Modular manufacturing process of the machine prototype. (**a**) Iron segments. (**b**) Segmented cell. (**c**) Single-cell wound. (**d**) Accomplished stator.

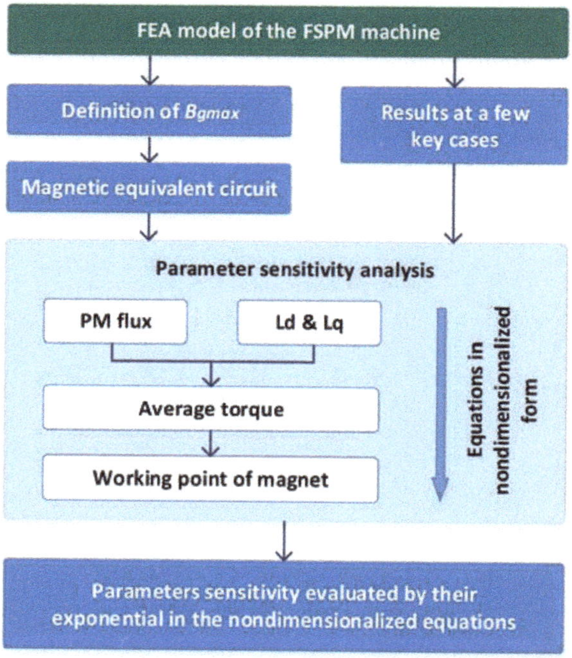

Figure 3. Flowchart of the parameter sensitivity analysis process.

2. Local Maximum Air-Gap Flux Density (B_{gmax}) and Magnetic Equivalent Circuit

First of all, the magnetic equivalent circuit of the FSPM machine is introduced. Only the rotor position where phase-A flux (Φ_{mA}) achieves peak value will be discussed, since it is directly related to the d-axis PM flux and thus the torque value. The MEC provides the bridge between the design parameters and electromagnetic performances. Figure 4 shows the flux distributions at d-axis when Φ_{mA} achieves peak value, and Figure 5 shows the corresponding flux density distributions along the air gap. As can be seen, the local maximum flux density appears where one rotor pole fully overlapped with the stator tooth in coil-A1, defined as B_{gmax}. Although the B_{gmax} is lower than the peak flux density along the air gap, as shown in Figure 5, it is more presentative than the absolute maximum value, since it is directly related to the phase-A flux linkage, and further the PM flux linkage of the FSPM motor.

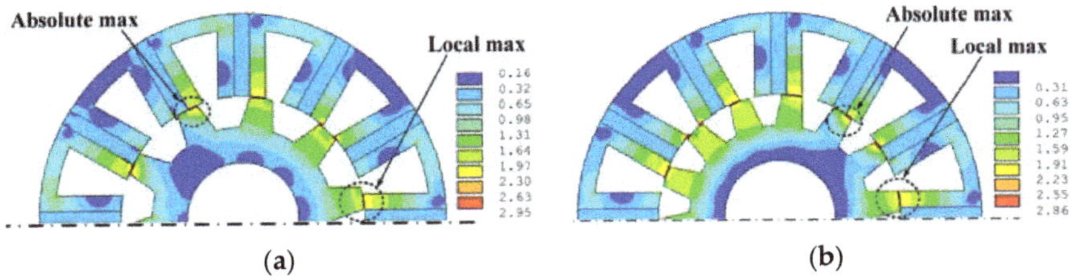

Figure 4. PM flux distributions when rotor is at d-axis. (**a**) 12/10 machine. (**b**) 12/14 machine.

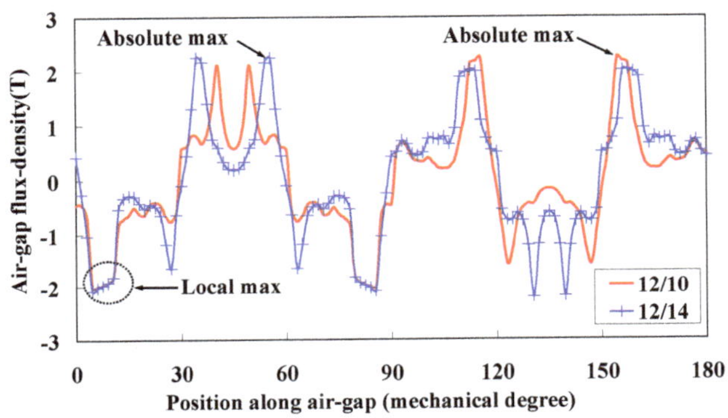

Figure 5. PM flux density distributions along air gap when rotor is at d-axis.

Based on the field distributions shown in Figure 4, a simplified MEC model for the FSPM machine can be obtained as shown in Figure 6. As can be seen, the air gap flux density at the position where the stator tooth surrounded by coil-A1 is fully overlapped with one rotor pole directly results in the maximum PM flux of coil-A1; thus, this will be used to calculate the flux linkages in coil-A1.

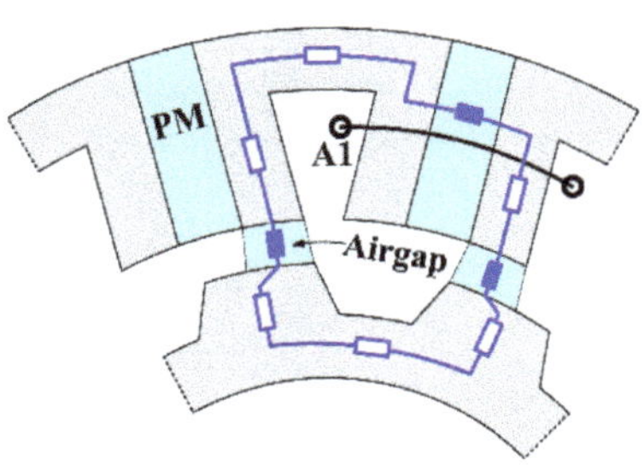

Figure 6. The MEC model and parameter specifications when rotor is at d-axis.

3. Investigation on the Parameter Sensitivities

Based on the MEC mode, the sensitivities of electromagnetic performances on parameters in the FSPM machine is studied in this part. Specifically, the influences of design parameters on the dominant electromagnetic characteristics, i.e., the PM flux, d-axis and q-axis inductances, average torque, power factor angle, and the working point of magnets are investigated. More attention is paid to the split ratio, i.e., the k_{sio} which is defined as $k_{sio} = D_{so}/D_{si}$, since it is directly related to the performances and more presentative.

3.1. PM Flux Linked by Armature Windings

When the B_{gmax} is defined, the PM flux in one stator pole Φ_p can be obtained by

$$\Phi_p = B_{gmax} w_{st} l_a \tag{1}$$

Φ_p is not fully linked by coil-A1, Figure 7 shows the PM flux leakages Φ_{leak} by different split ratios. Thus, the flux linked by coil-A1, Φ_{coil}, can be given by

$$\Phi_{coil} = \Phi_p - \Phi_{leak} \tag{2}$$

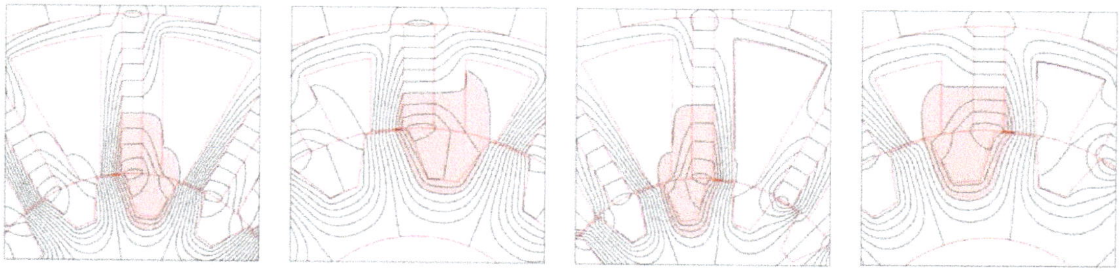

(a) (b) (c) (d)

Figure 7. Field distributions d-axis where the shadowed area refers to the leakage flux. (**a**) 12/10, k_{sio} = 0.5. (**b**) 12/10, k_{sio} = 0.7. (**c**) 12/14, k_{sio} = 0.5. (**d**) 12/14, k_{sio} = 0.7.

The flux leakage coefficient k_{flux} is introduced into Equation (1) to taking into account the leakage fluxes, then Φ_{coil} can be given by

$$\Phi_{coil} = \Phi_p k_{flux} = B_{gmax} w_{st} l_a k_{flux} \tag{3}$$

The PM flux of one phase Φ_m is given by

$$\Phi_m = \Phi_{coil} P_c \tag{4}$$

where P_c is the coil count per phase. Now the relationship between Φ_m and the key design parameters is built. To reveal the parameter sensitivities, the nondimensionalized form is introduced, which removes complex and tedious equations, but leaves the key characters in equations that present the correlation degree between the optimal goal and the parameters. The nondimensionalized form of Equation (4) can be given by

$$\begin{aligned} \Phi_m^* &= B_{gmax}^* w_{st}^* l_a^* k_{flux}^* P_c^* \\ &= \frac{B_{gmax}^* D_{so}^* k_{stw}^* k_{sio}^* l_a^* k_{flux}^* P_c^*}{P_s^*} \end{aligned} \tag{5}$$

Figure 7 also reveals that both the main PM flux and flux leakages increase with k_{sio}, and the square root relation between Φ_m and k_{sio} can be expected. The FEA results also indicates that, with satisfied accuracy, k^*_{flux} can be given by

$$k_{flux}^* = \sqrt{k_{sio}^*} \tag{6}$$

Thus, Equation (5) can be further expressed by

$$\Phi_m^* = \frac{B_{gmax}^* k_{stw}^* D_{so}^* l_a^* \left(k_{sio}^*\right)^{1.5} P_c^*}{P_s^*} \tag{7}$$

Equation (7) is validated by Figure 8 which compares the *d*-axis flux with good agreements where the base results for nondimensionalization is obtained at k_{sio} = 0.5. Equation (7) also reveals that the PM flux-linking armature windings is more sensitive to the split ratio than the other design dimensions.

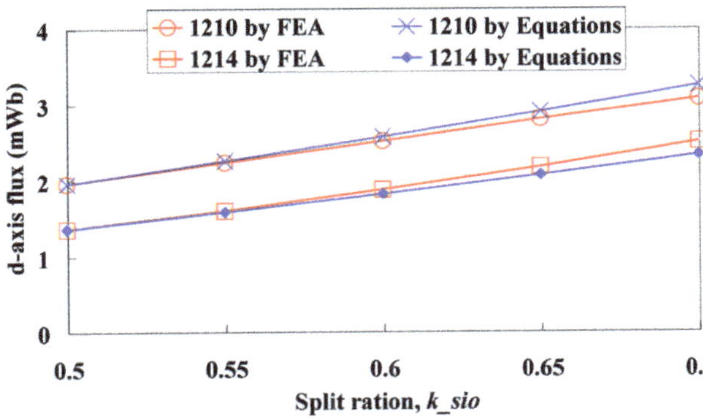

Figure 8. Comparison of the d-axis fluxes given by FEA and Equation (7).

3.2. d-Axis and q-Axis Inductances

The above analysis discussed the relationship of no-load performances with parameters. The study of loaded performances is started with investigation into the d-axis and q-axis inductances. First of all, Figures 9 and 10 present the field distributions in 12/10 and 12/14 machines when phase-A current is applied, where the blue lines indicate flux paths. The rotor positions at both the d-axis and q-axis are included. The blue solid lines indicate the path of the main fluxes, and the red dashed lines indicate the path of the leakage fluxes. As can be seen, when the rotor is at the d-axis, the permanent magnets wounded by coil-A1 and coil-A2 contribute the dominant magnetic reluctance of the main flux paths, which is decided by k_{pm}, k_{sio} and k_{st}. Additionally, when the rotor is at the q-axis, the air gap near the stator teeth of coil-A1 and coil-A2 provide the dominant magnetic reluctance of the main flux paths, which is also decided by k_{pm}, k_{sio} and k_{st}. With good accuracy, the nondimensionalized L_d, L_q can be given by Equation (8).

$$L_d^* = L_q^* = \frac{k_{st}^*}{k_{pm}^*} \cdot \frac{1}{k_{sio}^*} \tag{8}$$

Taking into account the stack length (l_a) and turns per coil (N_a), Equation (8) can be further presented by

$$L_d^* = L_q^* = l_a^* N_a^{*2} \frac{1}{k_{pm}^*} \cdot \frac{1}{k_{sio}^*} \tag{9}$$

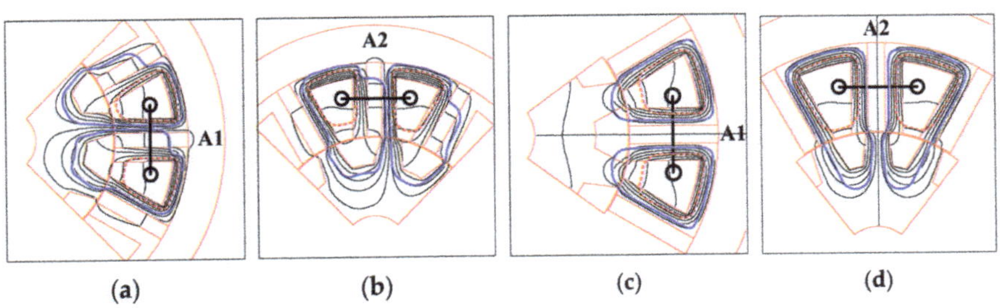

Figure 9. Field distributions in the 12/10 FSPM machine when only phase-A current is applied (**a**) coil-A1, at d-axis; (**b**) coil-A2, at d-axis; (**c**) coil-A1, at q-axis; (**d**) coil-A2, at q-axis.

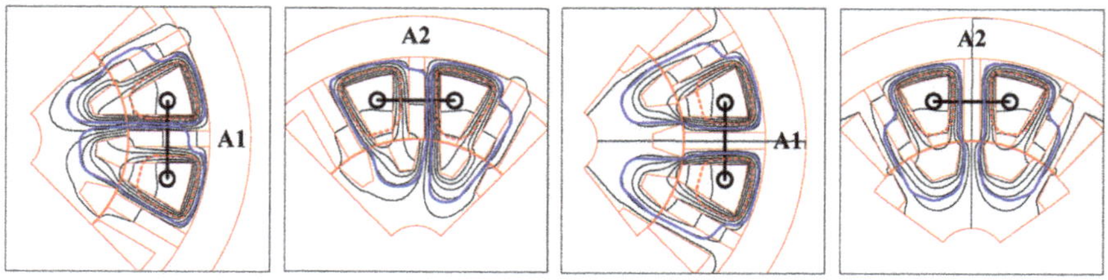

(a) (b) (c) (d)

Figure 10. Field distributions in the 12/14 FSPM machine when only phase-A current is applied: (**a**) coil-A1, at d-axis; (**b**) coil-A2, at d-axis; (**c**) coil-A1, at q-axis; (**d**) coil-A2, at q-axis.

Figure 11 compares the L_d and L_q calculated by FEA and Equation (9) in the 12/10 and 12/14 FSPM machines, where the base results for nondimensionalization is obtained at k_{sio} = 0.5. As can be seen, good agreement is achieved.

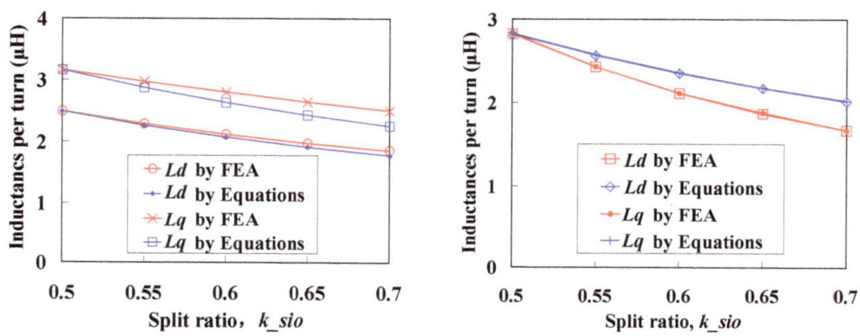

(a) (b)

Figure 11. The d-axis and q-axis inductances by FEA and Equation (9): (**a**) 12/10 machine; (**b**) 12/14 machine.

3.3. Electromagnetic Torque

The average torque will be discussed here. Usually, id = 0 control is adopted by the FSPM machines, and the electromagnetic torque T_e is calculated by

$$T_e = T_{pm} = \frac{3}{2}P_r\psi_m i_q = \frac{3}{2}P_r\psi_m i_s \quad (10)$$

Considering,

$$\begin{cases} \psi_m = \Phi_m N_a \\ i_s = i_q = \frac{\sqrt{2}\pi D_{so}k_{sio}}{P_s} \cdot \frac{A_s}{N_a} \end{cases} \quad (11)$$

where, N_{coil} and A_s are armature windings turns per coil and electrical load along air-gap (AT/m), A_s is given by

$$A_s = \frac{i_s N_a P_s}{\sqrt{2}\pi D_{so}k_{sio}}\Phi_m N_a \quad (12)$$

Bring Equations (11) and (12) into Equation (10), T_e can be given by

$$\begin{aligned} T_e = \tfrac{m}{2}P_r\psi_m i_q = \tfrac{m}{2}P_r(\Phi_m N_a)\left(\tfrac{\sqrt{2}\pi D_{so}k_{sio}}{P_s} \cdot \tfrac{A_s}{N_a}\right) \\ = \tfrac{m}{2} \cdot \tfrac{P_r}{P_s} \cdot \left(\Phi_m \sqrt{2}\pi D_{so}k_{sio}A_s\right) \end{aligned} \quad (13)$$

With acceptable accuracy, it can be considered $P_s/P_r \approx 1$ since it is normally $P_s = P_r \pm 2$ in the FSPM machines, so Equation (13) can be simplified as

$$T_e = \frac{m}{2}\left(\Phi_m \sqrt{2}\pi D_{so} k_{sio} A_s\right) \tag{14}$$

Then the nondimensionalized form T_e^* can be given by

$$T_e^* = m^* \Phi_m^* D_{so}^* k_{sio}^* A_s^* \tag{15}$$

Bringing Equation (7) into Equation (14), T_e^* can be presented by

$$\begin{aligned} T_e^* &= m^* \left(\frac{B_{gmax}^* k_{stw}^* D_{so}^* l_a^* \left(k_{sio}^*\right)^{1.5} P_c^*}{P_s^*} \right) D_{so}^* k_{sio}^* A_s^* \\ &= m^* B_{gmax}^* \left(k_{sio}^*\right)^{2.5} \left(D_{so}^*\right)^2 k_{stw}^* l_a^* A_s^* \end{aligned} \tag{16}$$

Equation (16) indicates that for a FSPM machine, the relationship between T_e and k_{sio} and A_s can be given by (keeping other parameters such as B^*_{gmax}, k^*_{stw} unchanged)

$$T_e^* = \left(k_{sio}^*\right)^{2.5} A_s^* \tag{17}$$

On the other hand, to keep the T_e unchanged with the variation of design parameter, A_s and k_{sio} should fit in Equation (18)

$$A_s^* = \left(k_{sio}^*\right)^{-2.5} \tag{18}$$

Figure 12 gives the torque waveform when A_s and k_{sio} fit in Equation (18). As can be seen, nearly the same average torque is achieved in the 12/10 and 12/14 machines, respectively, which also validates Equation (18).

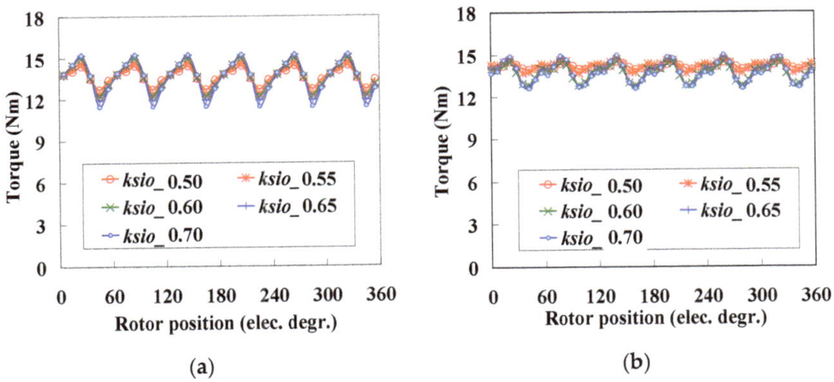

Figure 12. The torque waveforms when A_s and k_{sio} fit in Equation (18): (**a**) 12/10 machine; (**b**) 12/14 machine.

3.4. Power Factor Angle

The id = 0 control strategy is adopted in the FSPM machine due to a close L_d and L_q values, which inevitably reduces the power factor and increases the power factor angle. Thus in this part, the influences of design parameters on power factor angle is discussed. The vector frame is shown in Figure 13, and the tangent value of power factor angle φ can be given by

$$\tan\varphi = \frac{L_q i_q}{\psi_m N_a} = \frac{\Lambda_q N_a^2 i_s}{\Phi_m N_a} = \sqrt{2}\pi \frac{\Lambda_q D_{so} A_s k_{sio}}{\Phi_m P_s} \tag{19}$$

where Λ_q is the q-axis magnetic permeance. Then the nondimensionalized form, (tanφ)*, can be given by

$$(\tan\varphi)^* = \frac{\Lambda_q^* D_{so}^* A_s^*}{\Phi_m^*} = \frac{1}{B_{gmax}^*}\cdot\frac{A_s^*}{P_c^*}\cdot\frac{1}{k_{stw}^* k_{pm}^* k_{sio}^* k_{flux}^*} = \frac{1}{B_{gmax}^*}\cdot\frac{A_s^*}{P_c^*}\cdot\frac{1}{k_{pm}^*\left(k_{sio}^*\right)^{1.5}} \tag{20}$$

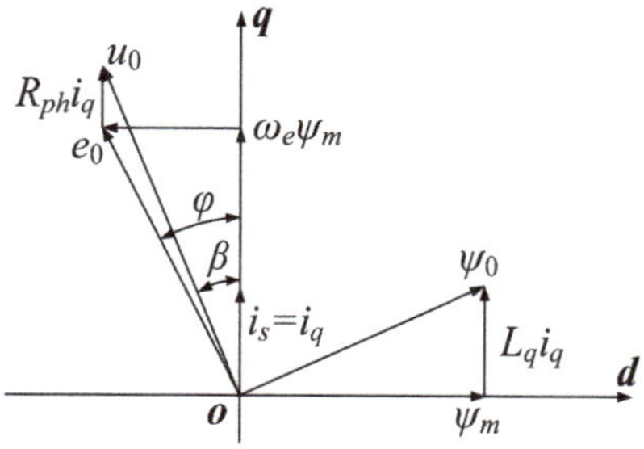

Figure 13. Vector frame employing $i_d = 0$ control strategy.

Equation (20) shows that using larger k_{pm} and k_{sio} will reduce tan φ, thus increasing the power factor, since larger k_{pm} and k_{sio} indicates a stronger PM field and a weaker armature reaction. Moreover, according to Equation (18), when the same torque is achieved the relationship between tan φ and k_{sio} can be given by bringing Equation (18) into Equation (20)

$$(\tan\varphi)^* = \frac{A_s^*}{\left(k_{sio}^*\right)^{1.5}} = \frac{1}{\left(k_{sio}^*\right)^{4}} \tag{21}$$

Figure 14 gives the variation of tan φ when the parameters fit in Equation (18) and, as can be seen, good agreements are achieved between FEA and Equation (21).

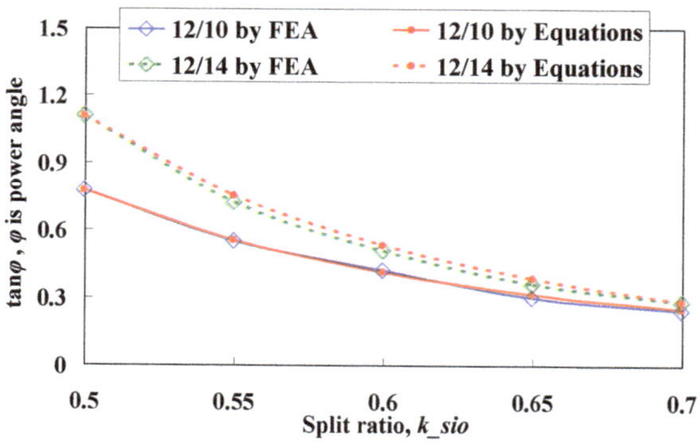

Figure 14. Comparison of the tan φ values given by FEA and Equation (21).

3.5. Working Point of the PMs

It is usually appreciated by a PM machine that the PMs works at the maximum energy product point. Thus, the working point of the PMs in the FSPM machine is discussed in this part. First of all, the working point of a PM (B_{pm}) is defined as

$$B_{pm} = \left[\frac{\sum_{i=1}^{max}\left(B_{elem-i}^2 S_{elem-i}\right)}{\sum_{i=1}^{max}\left(S_{elem-i}\right)}\right]^{\frac{1}{2}} \tag{22}$$

where $B_{elem\text{-}i}$ and $S_{elem\text{-}i}$ is the flux density and area of the *i*th element of the PM meshes in the FEA model. As can be seen from Figure 15, B_{pm} varies with the rotor position and k_{sio} since the magnetic circuit branches vary greatly.

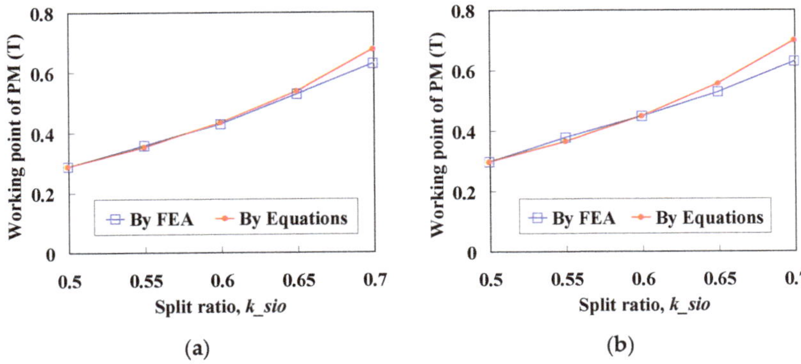

Figure 15. Comparison of the working point of PM (B_{pm}) due to different k_{sio}. (**a**) 12/10 machine. (**b**) 12/14 machine.

As shown in Figure 6, the flux generated by PM (Φ_{pm}) can be obtained by

$$\Phi_{pm} = \Phi_p = B_{gmax} w_{st} l_a \tag{23}$$

It should be noticed that the leakage flux is include in Φ_{pm}. Then, considering

$$\begin{cases} \Phi_{pm} = B_{pm} h_{pm} l_a \\ \Phi_p = B_{gmax} w_{st} l_a \end{cases} \tag{24}$$

It can be obtained that

$$B_{pm} h_{pm} l_a = B_{gmax} w_{st} l_a \tag{25}$$

$$B_{pm} = \frac{B_{gmax} w_{st} l_a}{h_{pm} l_a} = B_{gmax} \frac{\pi}{2P_s} \frac{k_{sio}}{1 - k_{sio}} \tag{26}$$

Then, the nondimensionalized form B^*_{pm} can be given by

$$(B_{pm})^* = \left(B_{gmax} \frac{\pi}{2P_s} \frac{k_{sio}}{1 - k_{sio}} \right)^* = \frac{B^*_{gmax}}{P^*_s} \cdot \frac{k_{sio}}{1 - k_{sio}} \tag{27}$$

Figure 15 gives the PM working point (B_{pm}) values due to different k_{sio}, where B_{pm} is the average value in one electrical cycle. As can be seen, good agreements are achieved between FEA results and Equation (21) predictions.

It should be emphasized that the proposed MEC method only covers the average performance in one electrical circle, e.g., the d-axis flux by magnets, average torque, etc., while the time-step-related performances such as torque ripples are not included, since only the rotor positions at d-axis and q-axis are considered.

4. Considerations Regarding Manufacturing Tolerances and Robust Approach

As mentioned in Section 1, the FSPM machine is a modular manufacturing method and the stator is combined by 12 segments. Although the modular stator facilitates the fabricating process and better copper fill, it will also generate large tolerances than traditional manufacturing methods, e.g., the uneven air gap length and concentricity errors, and leads to unbalanced radial forces and vibration. Thus, the influences of air-gap length on average and radial forces are studied in this section. Moreover, the robust design approach for the FSPM machine is briefly introduced, which aims to achieve best machine stability and robustness by setting boundaries on design dimensions, when taking into account the manufacturing tolerances.

4.1. Air-Gap Length and Key Performances

Inaccurate concentricity of the stator iron can hardly be avoided when assembling the modular segments into a housing/shell, which will lead to inaccurate air-gap lengths. Hence, a slightly smaller air-gap length might greatly increase the requirement of manufacturing accuracy and thus lead to higher costs during mass production. Therefore, special attention is paid to air-gap length firstly in this part. As can be seen from Figure 16 and Table 2, the torque output is reduced by only 9% when the air gap is increased by nearly 35~50%; meanwhile, the variations of loaded-phase EMF and cogging torque values can be neglected. This reveals that a larger air-gap length is preferred in the FSPM machine, since it can balance the slight drop of torque output and significantly reduce manufacturing difficulties and costs. In addition, a larger air gap will also reduce the unbalanced radial forces on rotor, as will be presented in next part.

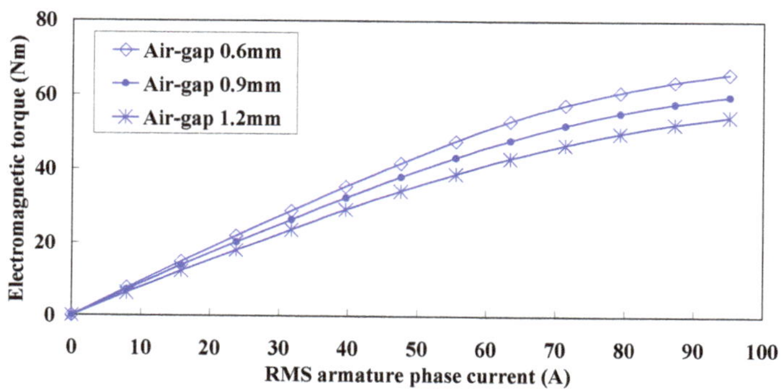

Figure 16. Comparison of the torque output by different air-gap lengths.

Table 2. Performances under different air-gap lengths by 2D FEA.

Electromagnetic Performances	Air-Gap Length		
	0.6 mm	0.9 mm	1.2 mm
RMS no-load phase-A EMF	38.0 V	35.3 V	30.8 V
Peak-to-peak cogging torque	4.8 Nm	3.3 Nm	3.2 Nm
RMS loaded phase-A EMF	56.8 V	55.7 V	54.9 V
Average torque, rated	66.3 Nm	60.5 Nm	55.1 Nm

4.2. Unbalanced Radial Forces on Rotor

The inaccurate stator concentricity might be generated by assembly tolerances, especially when modular segments are adopted. This will cause uneven air-gap lengths and unbalanced radial forces on the rotor, also called unbalanced magnetic pull (UMP). Although a precisely balanced rotor is preferred in these machines, it is hard to be obtained in integrated starter generator (ISG) systems in HEVs, where the rotor is directly coupled to the flywheel of the engine; consequently, the vibration of the engine may affect the concentricity of the electric motors. The unbalanced forces will lead to increased power loss, greater acoustic noise, vibration and thus degraded electrical isolations. Therefore, it is worth paying attention to this force. Based on 2D FEA, to simplify the analysis, the unbalanced forces are investigated by adopting an eccentric rotor with a bias of 0.3 mm from the stator center and towards coil-A1, as shown in Figure 17. Different air-gap lengths (under healthy conditions) are also considered.

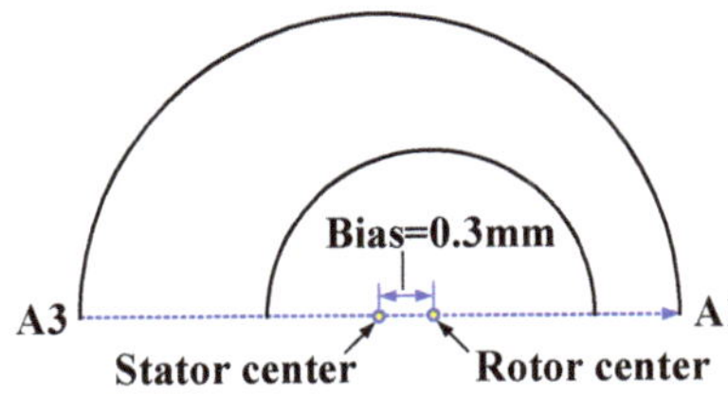

Figure 17. Illustration of the eccentric rotor with a bias from stator center and towards coil-A1.

Figures 18 and 19 give the unbalanced forces in one electrical period. Interestingly, the no-load unbalanced forces, i.e., due to PMs only, exhibit similar amplitudes (near 600 N) under different air-gap lengths, as can be seen in Figure 18, while the unbalanced forces under loaded conditions drop greatly when the air gap increases (from 820 N when g = 0.6 mm to 480 N when g = 1.2 mm), as shown in Figure 19. Meantime, the variations of other electromagnetic performances when the rotor bias occurs, e.g., winding inductances, back-EMF and electromagnetic torque waveforms, et al., can be neglected.

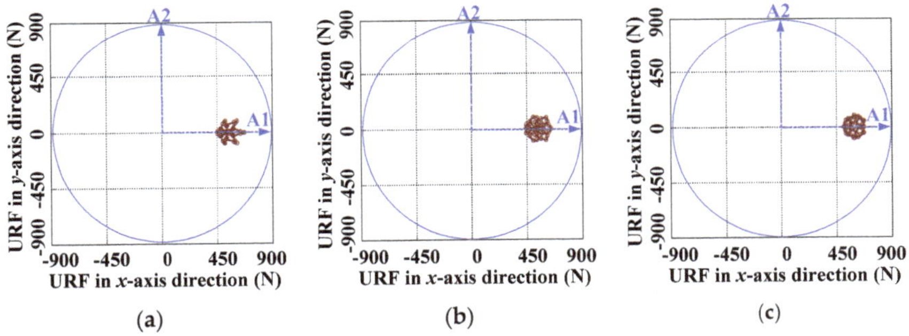

Figure 18. Unbalanced radial forces on rotor under no-load condition, when rotor bias is 0.3 mm. (**a**) Air-gap = 0.6 mm. (**b**) Air-gap = 0.9 mm. (**c**) Air-gap = 1.2 mm.

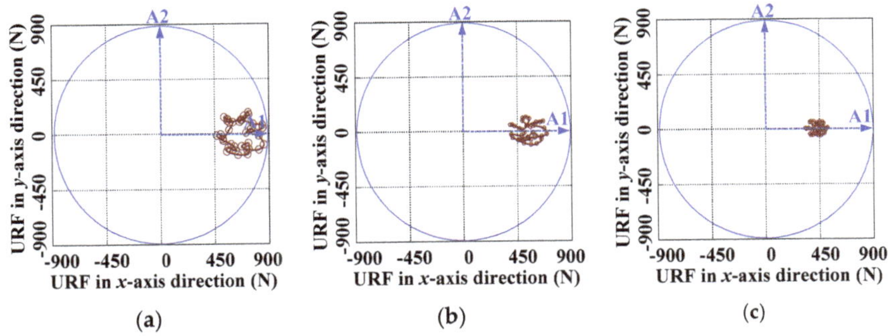

Figure 19. Unbalanced radial forces on rotor under loaded condition, when rotor bias is 0.3 mm. (**a**) Air-gap = 0.6 mm. (**b**) Air-gap = 0.9 mm. (**c**) Air-gap = 1.2 mm.

Overall, a relatively large air-gap length will contribute to the stability and robustness of the FSPM machine, thus a longer possible machine life, only with the acceptable drawback of slightly reduced torque output.

4.3. Robust Design Approach

Provided that manufacturing tolerances of dimensions follow the normal distribution, it can be derived that the electromagnetic and mechanical performances, e.g., the noise level, also follow the normal distribution. As can be seen from Figure 20a, the possibility of failed level, i.e., in the failure region can be significantly reduced in the optimized machine, by a robust design approach. However, this method may degrade the electromagnetic performances, since it aims at enhanced robustness, as shown in Figure 20b, where the regions R1 and R2 indicate the degraded electromagnetic capabilities and improved ones, respectively. Overall, maximizing region R2 while minimizing R1 is desired by the robust design approach. More details will be presented in a coming paper.

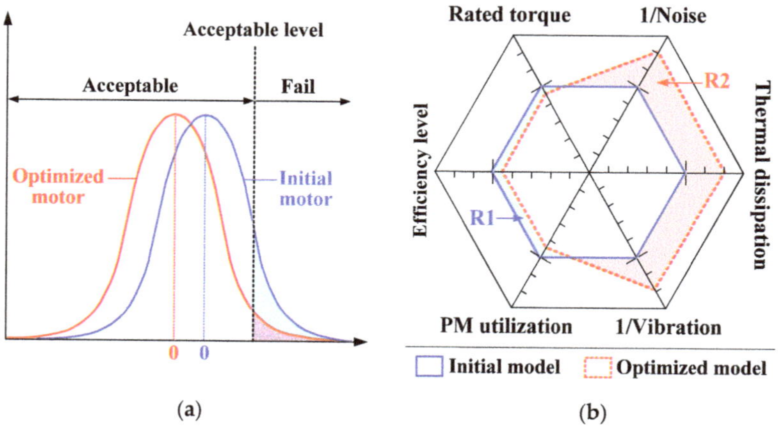

Figure 20. Illustration of the robust design approach for FSPM machine. (**a**) Normal distribution. (**b**) Graph of dimensions.

5. Experimental Validations and the Robust Design Approach

In this part, the experimental validations are carried out. Due to the relatively low ratios of stack length to stator outer diameter of the FSPM machine, the end-effect will cause considerable flux leakage along the axial direction, and hence, the 3D FEA predicted steady-state performances are compared to the experimental measurements as shown in Figure 21. As can be seen, good agreements are achieved, and due to saturation, the torque–current curve exhibits seriously nonlinear variation when the phase current exceeds 55A (RMS).

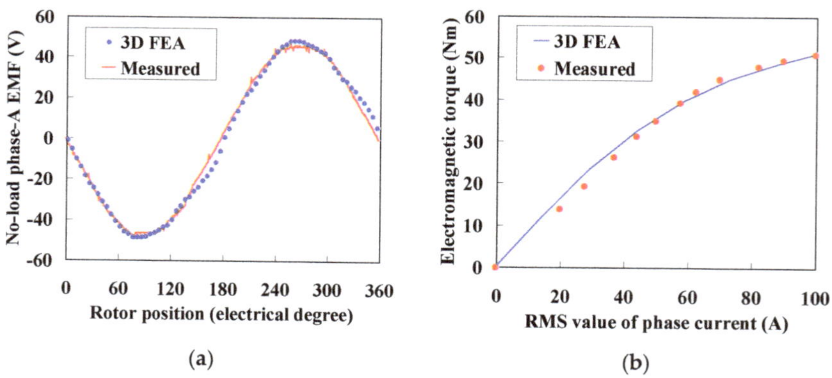

Figure 21. Comparison of the 3D FEA predicted and experimental measured performances. (**a**) No-load phase-A EMF waveforms at 1000 r/min. (**b**) Average electromagnetic torques.

6. Conclusions

Based on the magnetic equivalent magnetic circuit model, the parameter sensitivity analysis of the FSPM machine is carried out in this paper. Unlike the traditional methods, the proposed MEC-based method only needs FEA calculation of a few key cases, and thus is less time consuming, while obtaining satisfactory accuracy. The parameters' sensitivities are evaluated by their exponential in the nondimensionalized equations, and can be adopted to accelerate the multiobjective optimization of flux-switching machines in future work. It should be emphasized that the proposed MEC method only covers the average performance in one electrical circle, e.g., the average torque, PM flux, etc., while the time-step-related performances such as torque ripple are not included. Thereafter, focused on the influences of air-gap length on torque and unbalanced radial forces, the modular manufacturing method is discussed. Overall, a relatively large air-gap length will contribute to the stability and robustness of the FSPM machine. Finally, the robust design approach for the FSPM machine is introduced, which aims at best machine stability and robustness when taking into account the manufacturing tolerances. Experimental validations are also presented.

Author Contributions: Conceptualization, G.Z.; data curation, Q.T.; formal analysis, G.Z. and A.Q.; funding acquisition, Z.C.; investigation, G.Z., Q.T. and A.Q.; methodology, G.Z., Z.C. and X.X.; supervision, G.Z. and W.H.; writing—original draft, G.Z. All authors have read and agreed to the published version of the manuscript.

Funding: This work is supported in part by the National Nature Science Foundation of China under Grant 61973073, 52077032, in part by CALT under Grant CALT201806, in part by the Fundamental Research Funds for The Central Universities under Grant 2242021R41129, and in part by the Technology Foundation for Selected Overseas Chinese Scholar under Grant 1116000256.

Institutional Review Board Statement: Not applicable.

Informed Consent Statement: Not applicable.

Data Availability Statement: Not applicable.

Conflicts of Interest: The authors declare no conflict of interest.

References

1. Kurtović, H.; Hahn, I. Calculation of active material's torque contributions for a flux switching machine. *IEEE Trans. Magn.* **2020**, *56*, 7508504. [CrossRef]
2. Chen, Z.X.; Cui, Y.J. Numerical simulation and experimental validation of a flux switching permanent magnet memory machine. *IEEE Access* **2020**, *8*, 194904–194911. [CrossRef]
3. Hua, W.; Cheng, M.; Zhu, Z.Q. Analysis and optimization of back-EMF waveform of a flux-switching permanent magnet motor. *IEEE Trans. Energy Convers.* **2008**, *23*, 727–733. [CrossRef]
4. Cheng, M.; Hua, W.; Zhang, J.Z. Overview of stator-permanent magnet brushless machines. *IEEE Trans. Ind. Electron.* **2011**, *58*, 5087–5101. [CrossRef]
5. Chan, C.C. The state of the art of electric, hybrid, and fuel cell vehicles. *Proc. IEEE* **2007**, *95*, 704–718. [CrossRef]
6. Zhang, G.; Hua, W.; Cheng, M. Design and comparison of two six-phase hybrid-excited flux-switching machines for EV/HEV applications. *IEEE Trans. Ind. Electron.* **2016**, *63*, 481–493. [CrossRef]
7. Wang, P.X.; Hua, W.; Zhang, G.; Wang, B.; Cheng, M. Torque ripple suppression of flux-switching permanent magnet machine based on general air-gap field modulation theory. *IEEE Trans. Ind. Electron.* **2021**. Early access. [CrossRef]
8. Zeng, Z.Q.; Shen, Y.M.; Lu, Q.F.; Gerada, D.; Wu, B.; Huang, X.Y.; Gerada, C. Flux-density harmonics analysis of switched-flux permanent magnet machines. *IEEE Trans. Magn.* **2019**, *55*, 8103607. [CrossRef]
9. Wang, D.H.; Wang, X.H.; Jung, S.Y. Reduction on cogging torque in flux-switching permanent magnet machine by teeth notching schemes. *IEEE Trans. Magn.* **2012**, *48*, 4228–4231. [CrossRef]
10. Wang, P.X.; Hua, W.; Zhang, G.; Wang, B.; Cheng, M. Principle of flux-switching permanent magnet machine by magnetic field modulation theory part I: Back-electromotive-force generation. *IEEE Trans. Ind. Electron.* **2022**, *69*, 2370–2379. [CrossRef]
11. Wang, P.X.; Hua, W.; Zhang, G.; Wang, B.; Cheng, M. Principle of flux-switching pm machine by magnetic field modulation theory part II: Electromagnetic torque generation. *IEEE Trans. Ind. Electron.* **2022**, *69*, 2437–2446. [CrossRef]
12. Wang, P.X.; Hua, W.; Zhang, G.; Wang, B.; Cheng, M. Inductance characteristics of flux-switching permanent magnet machine based on general air-gap filed modulation theory. *IEEE Trans. Ind. Electron.* **2021**. Early access. [CrossRef]
13. Zhu, Z.Q.; Pang, Y.; Howe, D. Analysis of electromagnetic performance of flux-switching permanent-magnet machines by nonlinear adaptive lumped parameter magnetic circuit model. *IEEE Trans. Magn.* **2005**, *41*, 4277–4287. [CrossRef]
14. Ilhan, E.; Gysen, B.L.J.; Paulides, J.J.H. Analytical hybrid model for flux switching permanent magnet machines. *IEEE Trans. Magn.* **2010**, *46*, 1762–1765. [CrossRef]

15. Gysen, B.L.J.; Ilhan, E.; Meessen, K.J. Modeling of flux switching permanent magnet machines with Fourier analysis. *IEEE Trans. Magn.* **2010**, *46*, 1499–1502. [CrossRef]
16. Zhu, S.; Cheng, M.; Hua, W.; Cai, X.H.; Tong, M.H. Finite element analysis of flux-switching pm machine considering oversaturation and irreversible demagnetization. *IEEE Trans. Magn.* **2015**, *51*, 7403404. [CrossRef]
17. Sulaiman, E.; Kosaka, T. Parameter sensitivity study for optimization of field-excitation flux switching synchronous machine for hybrid electric vehicles. In Proceedings of the 2012 7th IEEE Conference on Industrial Electronics and Applications (ICIEA), Singapore, 18–20 July 2012. [CrossRef]
18. Xiang, Z.X.; Zhu, X.Y.; Quan, L.; Fan, D.Y. Optimization design and analysis of a hybrid permanent magnet flux-switching motor with compound rotor configuration. *CES Transact. Electr. Mach. Syst.* **2018**, *2*, 200–206. [CrossRef]
19. Chen, Y.Y.; Zhuang, J.H.; Ding, Y.; Li, X.J. Optimal design and performance analysis of double stator multi-excitation flux-switching machine. *IEEE Trans. Appl. Supercond.* **2019**, *2*, 0601205. [CrossRef]
20. Yu, J.C.; Liu, C.H. Multi-objective optimization of a double-stator hybrid-excited flux-switching permanent-magnet machine. *IEEE Trans. Energy Convers.* **2020**, *35*, 312–323. [CrossRef]
21. Ullah, W.; Khan, F.; Umair, M. Multi-objective optimization of high torque density segmented PM consequent pole flux switching machine with flux bridge. *CES Transact. Electr.Mach. Syst.* **2021**, *5*, 30–40. [CrossRef]
22. Sun, H.; Yang, J. Optimal decisions for competitive manufacturers under carbon tax and cap-and-trade policies. *Comput. Ind. Eng.* **2021**, *156*, 107244. [CrossRef]
23. Foumani, M.; Jenab, K. Cycle time analysis in reentrant robotic cells with swap ability. *Int. J. Product. Res.* **2012**, *50*, 6372–6387. [CrossRef]
24. Sulaiman, E.; Zakaria, S.N.U.; Kosaka, T. Parameter sensitivity study for optimization of single phase E-Core hybrid excitation flux switching machine. In Proceedings of the 2015 IEEE International Conference on Mechatronics (ICM), Nagoya, Japan, 6–8 March 2015. [CrossRef]
25. Zhao, W.J.; Qiu, X.; Bu, F.F.; Yang, J.F.; Fan, W.X. Study on influence of structure parameters on electromagnetic characteristics for 12/10 flux-switching permanent magnet motor. In Proceedings of the 2019 IEEE 13th International Conference on Power Electronics and Drive Systems (PEDS), Toulouse, France, 9–12 July 2019. [CrossRef]
26. Lei, G.; Wang, T.S.; Zhu, J.G. System-level design optimization method for electrical drive systems—robust approach. *IEEE Trans. Ind. Electron.* **2015**, *62*, 4702–4713. [CrossRef]

energies

Article

Thermal Model Approach to the YASA Machine for In-Wheel Traction Applications

Guangchen Wang [1], Yingjie Wang [2], Yuan Gao [3], Wei Hua [4], Qinan Ni [5] and Hengliang Zhang [4,*]

1 Taizhou Liangsu Technology Co., Ltd., Taizhou 318016, China; m15909828335@163.com
2 Tangshan Power Supply Company, State Grid Jibei Power Co., Ltd., Tangshan 063000, China; yjwang1995@163.com
3 Department of Aerospace Engineering, Faculty of Engineering, University of Bristol, Queens Road, Bristol BS8 1QU, UK; yuan21.gao@bristol.ac.uk
4 School of Electrical Engineering, Southeast University, Nanjing 210096, China; huawei1978@seu.edu.cn
5 Jiangsu Yueda Group, Yancheng 224007, China; niqn_hit@163.com
* Correspondence: zhanghengliang@seu.edu.cn

Abstract: The axial-flux permanent magnet (AFPM) machines with yokeless and segmented armature (YASA) topology are suitable for in-wheel traction systems due to the high power density and efficiency. To guarantee the reliable operation of the YASA machines, an accurate thermal analysis should be undertaken in detail during the electrical machine design phase. The technical contribution of this paper is to establish a detailed thermal analysis model of the YASA machine by the lumped parameter thermal network (LPTN) method. Compared with the computational fluid dynamics (CFD) method and the finite element (FE) method, the LPTN method can obtain an accurate temperature distribution with low time consumption. Firstly, the LPTN model of each component of the YASA machine is constructed with technical details. Secondly, the losses of the YASA machine are obtained by the electromagnetic FE analysis. Then, the temperature distribution of the machine can be calculated by the LPTN model and loss information. Finally, a prototype of the YASA machine is manufactured and its temperature distribution under different operating conditions is tested by TT-K-30 thermocouple temperature sensors. The experimental data matches the LPTN results well.

Keywords: lumped parameter thermal network (LPTN); losses; temperature distribution; yokeless and segmented armature (YASA)

Citation: Wang, G.; Wang, Y.; Gao, Y.; Hua, W.; Ni, Q.; Zhang, H. Thermal Model Approach to the YASA Machine for In-Wheel Traction Applications. *Energies* **2022**, *15*, 5431. https://doi.org/10.3390/en15155431

Academic Editor: Federico Barrero

Received: 7 July 2022
Accepted: 25 July 2022
Published: 27 July 2022

1. Introduction

Electric vehicles driven by four in-wheel electrical machines make the traction system more simplified and flexible [1]. In the limited volume, the yokeless and segmented armature (YASA) axial flux machine is capable of producing larger torque compared with radial flux machines [2–5]. The yokeless segmented stator core and the centralized short-distance windings make it more efficient and give it a higher power density [6,7]. Therefore, YASA machines are potential candidates for in-wheel traction applications.

It is well-known that losses cause the temperature rise of electrical machines, which can weaken the machine's performance and even damage electrical machines [8]. The insulation system of armature winding and thermal characteristics of the permanent magnet limits the maximum temperature of the electrical machines. The YASA machines feature a high power density, which means a high loss density as well [9,10]. When machines produce the peak torque under the peak current, a high copper loss is generated, making the winding temperature rise rapidly. That challenges the insulation system of the electrical machine. To ensure the reliability of the YASA machine, the thermal performance should be thoroughly considered [11,12].

There are three typical methods for the thermal analysis of YASA machines, i.e., the computational fluid dynamics (CFD) method, the finite element (FE) method, and the lumped parameter thermal network (LPTN) method [13–21]. The CFD method can calculate the convective heat transfer coefficients, which are significant in electrical machine design [1,13]. However, the CFD model is usually complicated, which means it requires several days or even a week to establish and calculate the thermal model [14]. In [15], a novel stator cooling structure is introduced to improve the temperature distribution of the YASA machine, and the CFD method verifies the proposed cooling structure. The CFD method is applied in [16] to calculate an innovative water-cooling system for the YASA machine. The FE method can obtain an accurate temperature distribution, but it requires several hours to calculate the model and the boundary processing is complicated and difficult [17,18]. In [17], both analytical and experimental investigations into the air-cooling of a YASA motor for in-wheel traction are presented. In [18], a stator heat extraction system for YASA machines is introduced and modeled, and the thermal analysis is calculated by the FE method. The LPTN method divides the electrical machine into several nodes corresponding to the machine components and the heat-source distribution [19,20], which can guarantee the analysis accuracy while keeping it fast compared with the CFD and FE methods. For machines with complex structures, accurate temperature distribution can be quickly obtained by LPTN. Therefore, this paper only focuses on a LPTN model of the YASA machine.

To the best of our knowledge, only a few papers have used the LPTN method to compute the temperature distribution of YASA machines. For example, in [21], the short-time duty and the intermittent duty of a 4 kW YASA machine are calculated based on the LPTN method. In [22], a lumped parameter T-type thermal network model is applied to the YASA machine temperature calculation, but it mainly focuses on the calculation of core loss rather than the total losses. In [23], a 3D LPTN model is developed and experimentally validated, where the air-cooling channels between permanent magnets on the rotor plates are also considered. However, it only constructs the thermal models of the stator (including the armature winding and stator core) and rotor without considering the other components of the YASA machine. The main technical contribution of this paper is to use the LPTN method to construct a detailed thermal model for the YASA machine including all components. The temperature distribution of the YASA machine is obtained rapidly and accurately by considering the losses calculated by the FE method. The LPTN model can provide a reference for the thermal analysis of YASA machines, which not only ensures the thermal reliability of the machine, but also saves time and calculation resources in the machine design phase.

This report is arranged as follows: In Section 2, the topology of the YASA machine is introduced, and the corresponding design parameters are listed. Then, in Section 3, the detailed thermal model of the YASA machine is established, and the corresponding thermal resistance of all components is deduced. To obtain the temperature distribution, the calculated losses are introduced into the entire thermal model of the YASA machine, as described in Section 4. After that, in Section 5, the prototype test proves the accuracy of the thermal network model. Finally, this paper is concluded in Section 6.

2. Topology of the Studied YASA Machine

In this section, the structure and design parameters of the YASA machine are introduced.

2.1. Structure

Similar to the radial-flux permanent magnet (RFPM) machines, the YASA machines belong to the category of permanent magnet (PM) machines. However, the magnetic flux in the air gap of YASA machines is different from that of the RFPM machines. The magnetic flux of the YASA machines is along the axial direction in the air gap, while that of the RFPM machines is along the radial direction. Compared with the conventional RFPM machines, the YASA machines can generate a higher performance when the radial diameter is larger than the axial length [1,8,17].

The YASA machines consist of a yokeless and segmented stator and double external rotors [8]. The main magnetic flux of YASA machines, shown in Figure 1, starts from the N-pole permanent magnet and passes through the stator core to the S-pole on the other side. After passing through the rotor core on the second side, the flux starts from the N-pole on the second side and passes through the stator core to the S-pole of the first side. Finally, the main magnetic flux forms a closed loop.

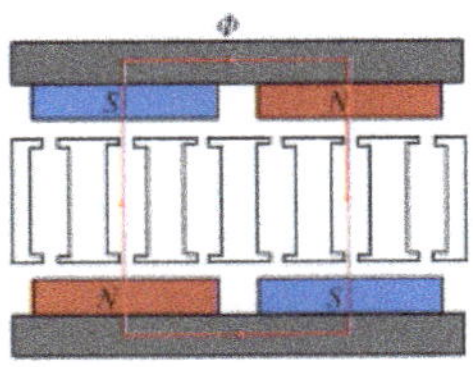

Figure 1. The main magnetic flux of the YASA machines.

The construction of YASA machines possesses many significant advantages. The segmented stator core equipped with a concentrated winding having short end-windings that result in a high filling factor and low copper losses. The structure of a yokeless stator is beneficial to decreasing mass and core loss [1,8,17]. In addition, YASA machines generally exhibit low self-inductance and mutual inductances among the phases, which improve fault tolerance and operation reliability [16,19].

The YASA machine investigated in this paper is shown in Figure 2, where the x-axis, y-axis, and z-axis, respectively, represent the circumferential direction, axial direction, and radical direction. The YASA machine can be divided into the stationary part and rotating part, where the stationary part includes the stator core, armature winding, supporting frame, and shaft, and the rotating part includes permanent magnets, rotor core, and the housing.

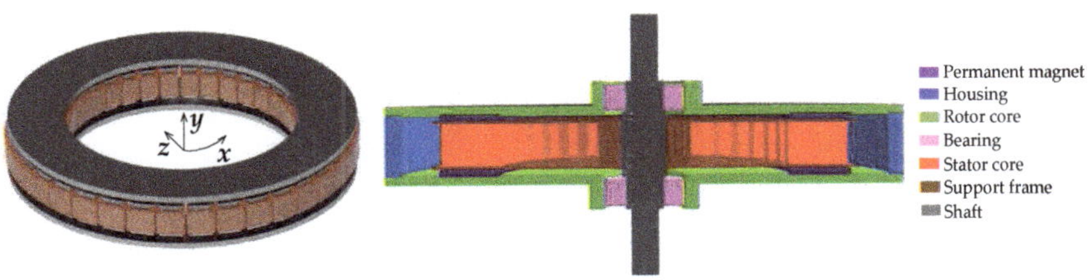

(a) (b)

Figure 2. Structure of the YASA machine: (**a**) 3-dimensional structure; (**b**) view of the cross-section.

2.2. Design Parameters

A 5 kW prototyped machine was developed to validate the thermal analysis of the YASA machine. The main performance and geometric parameters are listed in Table 1.

Table 1. Key parameters of the studied YASA machine.

Parameters	Values	Parameters	Values
DC voltage (V)	72	Active outer diameter (mm)	270
Rated power (kW)	5	Active internal diameter (mm)	190
Rated speed (r/min)	480	Active axial length (mm)	45
Rated torque (Nm)	99	Rotor core thickness (mm)	4.8
Rated current (Arms)	17	Permanent magnet thickness (mm)	3.4
Slot number	36	Pole arc coefficient	0.83
Pole number	32	Air gap (mm)	0.9

3. Construction of the LPTN Model

In this section, the LPTN models of different components of the YASA machine are constructed. Generally, there are three main parts taken into account, i.e., the stationary part, the rotating part, and the YASA machine assembly. The thermal model of the YASA machine involves many parameters (including geometric parameters and heat transfer coefficients). The geometric parameters of the YASA machine are determined by the intelligent optimization algorithm in the machine design process. Additionally, the heat transfer coefficients are identified by fine-tuning against simple test data through a genetic algorithm [24].

3.1. Thermal Model of the Stationary Part

In this subsection, the thermal model of the stationary part is introduced, which includes the armature winding, stator core, support frame, and shaft.

3.1.1. Armature Winding

The armature winding is the main heat source, especially when operating at peak torque with peak current. The heat dissipation of the armature winding is difficult, because the winding is in the middle of the YASA machine. Additionally, the winding is relatively dispersed, resulting in uneven temperature distribution. Hence, it is difficult to build a detailed thermal model.

This paper uses the layered winding model from [25], in which the coil of the armature winding is regarded as a uniform heat conduction material. The actual thermal conductivity is replaced by an equivalent thermal conductivity k_{wd}.

The armature winding includes the outer end-winding, inner end-winding, and slot-winding. Figure 3 shows the specific position of the armature winding. Each section of the armature winding is represented in Figure 4. The numbers in Figure 4 correspond to those marked in Figure 3.

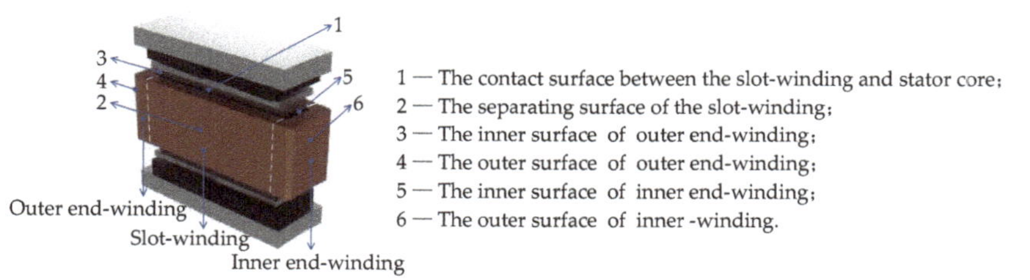

Figure 3. Position in the cross-section of the segmented stator.

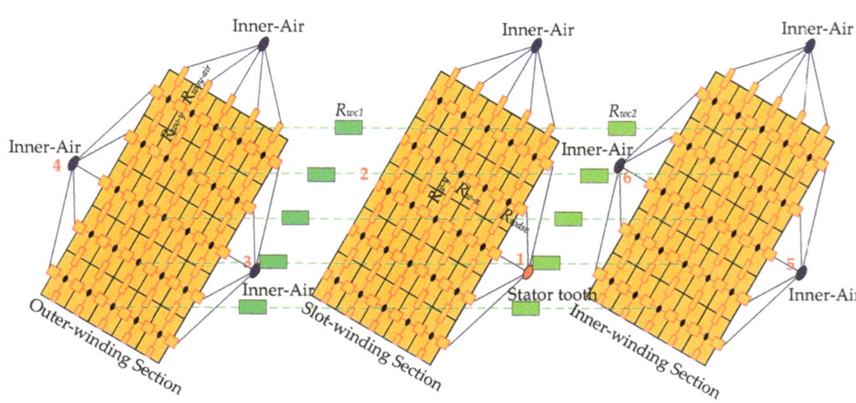

Figure 4. The thermal model of a single segmented armature winding.

For both the inner and outer end-winding, axial thermal resistances (R_{wo-y}, $R_{woy-air}$, R_{wo-ya}, R_{w-y}) and radial thermal resistance (R_{w-x}) are considered. Since the axial cross-section of the end-windings remains unchanged, the axial thermal resistances are approximately constant. The radial thermal resistance changes with the radius. For instance, the axial thermal resistance of the outer end-winding is

$$R_{wo-y} = \frac{L_{wo-y}}{k_{wd} S_{wo-y}} \tag{1}$$

where L_{wo-y} is the distance of adjacent nodes at the outer end-winding, and S_{wo-y} is the area of the corresponding region between nodes, as shown in Figure 4.

Between the outer end-winding and inner-air (including the air gap and air inside the YASA machine), there are axial thermal resistances $R_{woy-air}$ and R_{wo-ya}.

$$R_{woy-air} = \frac{1}{2} R_{wo-y} + R_{wo-ya} \tag{2}$$

$$R_{wo-ya} = \frac{1}{k_{wdout} S_{wo-y}} \tag{3}$$

where R_{wo-ya} is the convective thermal resistance and k_{wdout} is the equivalent convective heat transfer coefficient.

Similarly, the cross-section of slot-winding is constant. The thermal resistances of the slot-winding (including the axial thermal resistance R_{w-y} and radial thermal resistance R_{w-x}) are regarded as unchanged. The expressions follow:

$$R_{w-y} = \frac{\frac{H_{wd}}{2}/5}{\left(\frac{W_{wd}}{2}/5 \times L_s\right) \times k_{wd}} \tag{4}$$

$$R_{w-x} = \frac{\frac{W_{wd}}{2}/5}{\left(\frac{H_{wd}}{2}/5 \times L_s\right) \times k_{wd}} \tag{5}$$

where W_{wd} and H_{wd} represent the width and height of slot-winding, respectively. L_s is the difference between the outer and inner radius of the stator core.

When calculating the thermal resistance R_{wdsc} between the slot-winding and stator core, the thermal resistance of slot insulation should be taken into account:

$$R_{wdsc} = \frac{1}{2} R_{w-x} + R_{lx-s} + R_{scx} \tag{6}$$

$$R_{lx-s} = \frac{W_l}{\left(\frac{H_{wd}}{2}/5 \times L_s\right) \times k_l} \tag{7}$$

where R_{lx-s} is the thermal resistance of the slot insulation, W_l is the equivalent thickness of the slot insulation, k_l is the equivalent thermal conductivity of the slot insulation, and Rs_{cx} is the thermal resistance of the stator core along the x-axis. In addition, there are thermal resistances R_{wc1} and R_{wc2} between the slot-winding and both end-windings, respectively, as shown in Figure 4.

3.1.2. Stator Core

The complete thermal model of the stator is shown in Figure 5. It should be noted that the thermal model of the armature winding and segmented stator core can be consolidated into a single node [19]. R_{ciax} and R_{ciay} are the equivalent thermal resistance of the armature winding along the x-axis and y-axis, respectively. R_{lx} is the thermal resistance of the slot insulation. R_{wia} is the convective thermal resistance between the armature winding and inner-air. $P_{w,slot}$ and $C_{w,slot}$ represent the loss and heat capacity of the 1/4 slots, respectively.

$P_{sc,slot}$ and $C_{sc,slot}$ represent the loss and heat capacity of the 1/2 stator core, respectively. The heat capacity is defined as follows:

$$C_x = c_x m_x \tag{8}$$

where m_x and c_x are the mass and heat capacity, respectively.

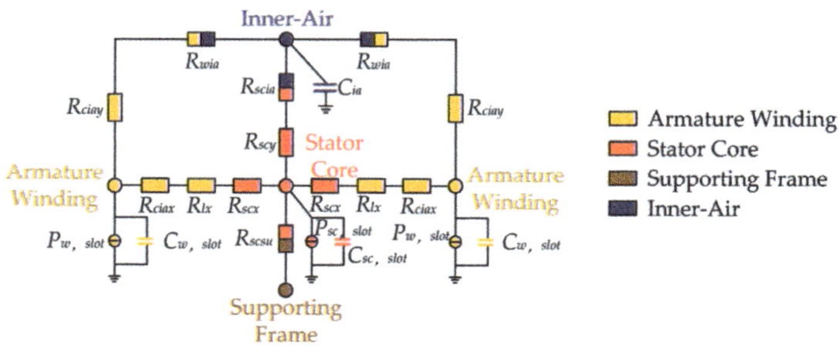

Figure 5. Thermal model of the single-segmented stator.

R_{scx} and R_{scy} are the thermal resistances of the stator core along the x-axis and y-axis, respectively. The calculations are as follows:

$$R_{scx} = \frac{W_{st}/2}{\frac{H_{st}}{2} \times L_s k_{iron}} \tag{9}$$

$$R_{scy} = \frac{H_{st}/4}{W_{st} L_s k_{iron}} \tag{10}$$

where W_{st} and H_{st} are the width and height of the stator core, respectively, and k_{iron} is the thermal conductivity of the stator core.

Owing to the symmetrical structure, Figure 5 can be further simplified to Figure 6. Figure 6 is the thermal model of the stator core, which constitutes the entire thermal model of the YASA machine (discussed in Section 3.3). P_w and C_w represent half of the copper loss and heat capacity. P_{sc} and C_{sc} represent half of the stator core loss and heat capacity. The convective thermal resistance between the stator core and inner-air present is R_{scia}.

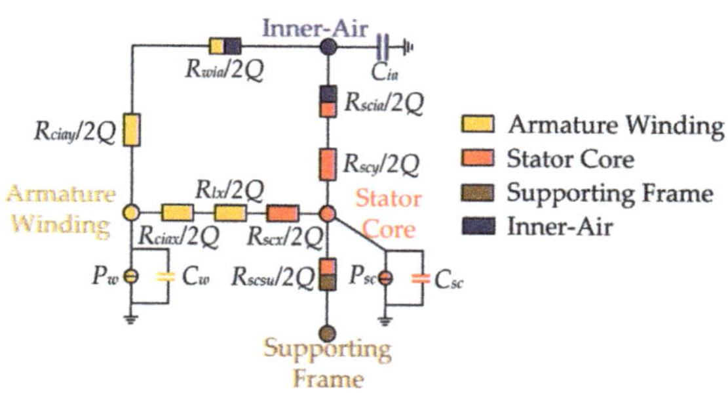

Figure 6. Simplified thermal model of the segmented stator.

3.1.3. Support Frame

The segmented stator core is shown in Figure 7a. In Figure 7b, there are matched cages on both sides of the segmented stator cores for support. The cages are connected by a circular connector with a support frame.

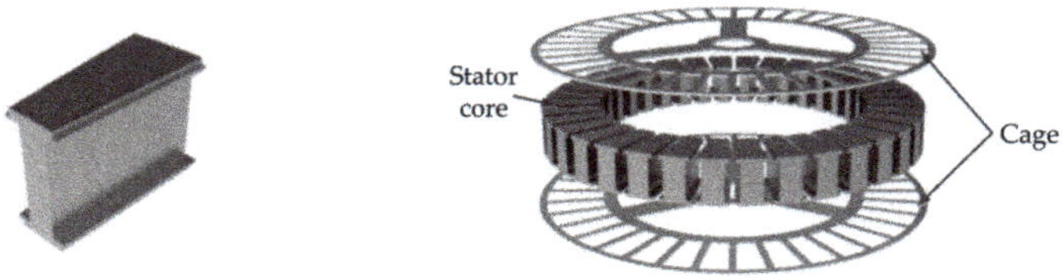

(a) (b)

Figure 7. Stator: (**a**) segmented stator core; (**b**) stator core and cages.

The thermal model of the support frame is shown in Figure 8. The support frame connects with the stator core and shaft. The thermal resistance between the support frame and stator core is R_{scsu}, as shown in Figure 6. The thermal resistance between the support frame and shaft is R_{sush}. In addition, there is thermal resistance R_{suia} between the support frame and inner-air. The calculation equations follow:

$$R_{scsu} = \frac{l_{scsu}}{S_{scsu}k_{air}} \tag{11}$$

$$R_{sush} = \frac{l_{sush}}{S_{sush}k_{air}} \tag{12}$$

$$S_{sush} = 2\pi R_{sui}\frac{H_{su}}{2} = \pi R_{sui}H_{su} \tag{13}$$

where l_{scsu} is the tolerance clearance between the support frame and stator core, l_{sush} is the tolerance clearance between the support frame and shaft, S_{scsu} is the area of the tolerance clearance between the support frame and stator core, S_{sush} is the area of the tolerance clearance between the support frame and shaft, and k_{air} is the thermal conductivity of air. R_{sui} and H_{su} are the inner radius and height of the connector.

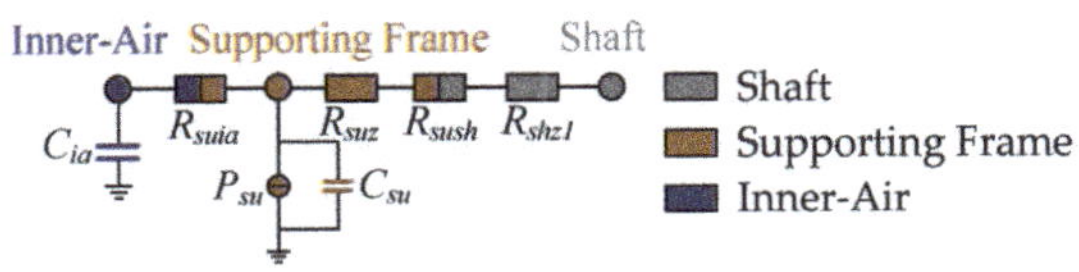

Figure 8. The thermal model of the support frame.

3.1.4. Shaft

Figure 9 shows the thermal model of shaft, where R_{shz1} represents the radial thermal resistance from the support frame to shaft, R_{shz2} is the radial thermal resistance from the shaft to bearing, R_{shy1} is the axial thermal resistance between the shaft and bearing, R_{shy2} is the axial thermal resistance from the bearing to shaft end, R_{shso} is convective thermal resistance between the shaft and surroundings. Additionally, there is convection thermal resistance R_{shia} between the shaft and inner-air. The thermal resistances are calculated as follows:

$$R_{shz1} = \frac{d_{sh1}/2}{(\pi d_{sh1}/2)\frac{H_{sh}}{2}k_{sh}} = \frac{2}{\pi H_{sh}k_{sh}} \tag{14}$$

$$R_{shz2} = \frac{d_{sh2}/2}{(\pi d_{sh2}/2)H_{be}k_{sh}} = \frac{1}{\pi H_{be}k_{sh}} \tag{15}$$

$$R_{shy1} = \frac{l_{sh1}/4}{\pi(d_{sh1}/2)^2k_{sh}} + \frac{l_{sh2}/2}{\pi(d_{sh2}/2)^2k_{sh}} \tag{16}$$

$$R_{shy2} = \frac{l_{sh2}/2}{\pi(d_{sh2}/2)^2k_{sh}} + \frac{l_{sh3}/2}{\pi(d_{sh3}/2)^2k_{sh}} \tag{17}$$

$$R_{shso} = \frac{1}{S_{shso}k_{shso}} \tag{18}$$

$$S_{shso} = \pi(d_{sh3}/2)^2 + \pi d_{sh3}l_{sh3} \tag{19}$$

where k_{sh} is the thermal conductivity of the shaft, k_{shso} is the convective heat transfer coefficient between the shaft and surroundings, and H_{be} is the height of the bearing. The other structural parameters are defined in Figure 10.

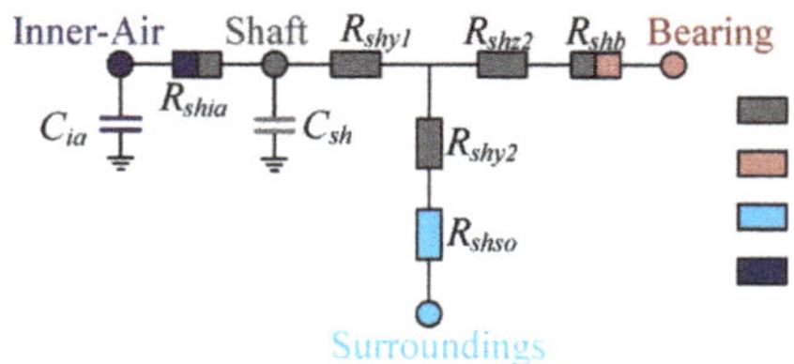

Figure 9. Thermal model of the shaft.

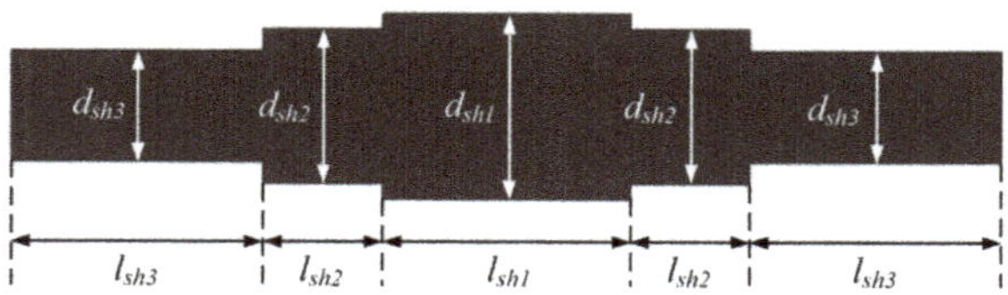

Figure 10. Structure of the shaft.

3.2. Thermal Model of the Rotating Part

In this subsection, the thermal model of the rotating part is introduced, which includes the permanent magnets, bearing, rotor core, and housing.

3.2.1. Permanent Magnet

Figure 11 shows the thermal model of the permanent magnet, where R_{mia} is the convective thermal resistance between the permanent magnet and inner-air, R_{my} is the axial thermal resistance of the permanent magnet, R_{mrc} is the thermal resistance between the permanent magnet and rotor core, and R_{rcy} is the axial thermal resistance of rotor core.

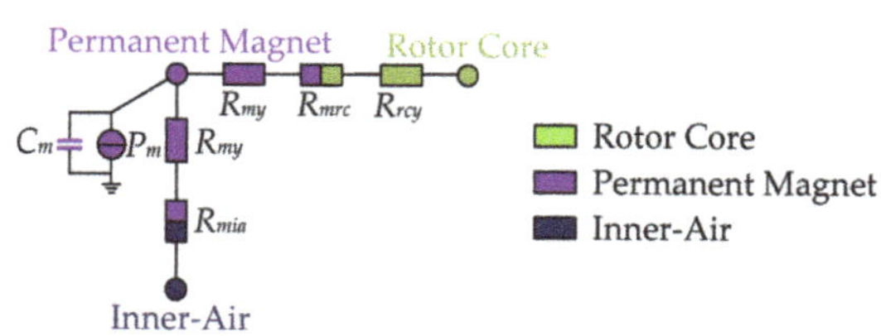

Figure 11. Thermal model of the permanent magnet.

The equations of R_{my}, R_{mcr} and R_{rcy} follow:

$$R_{my} = \frac{H_m/2}{S_m k_m} \tag{20}$$

$$R_{mrc} = \frac{l_{mr}}{S_m k_{air}} \tag{21}$$

$$R_{rcy} = \frac{H_{rc}/2}{S_{rc}k_{rc}} \tag{22}$$

$$S_{rc} = \pi\left(r_{rco}^2 - r_{rci}^2\right) \tag{23}$$

where H_m and H_{rc} are the thicknesses of the permanent magnet and rotor core, respectively. S_m, S_{rc} is the radial cross-section area of the permanent magnet and rotor core, respectively; k_m, k_{rc} is the thermal conductivity of permanent magnet and rotor core, respectively. The tolerance clearance between the permanent magnet and rotor core is l_{mr}; r_{rco}, r_{rci} is the outer and inner radius of the rotor core, respectively. The components of the machine cross-section are illustrated in Figure 12.

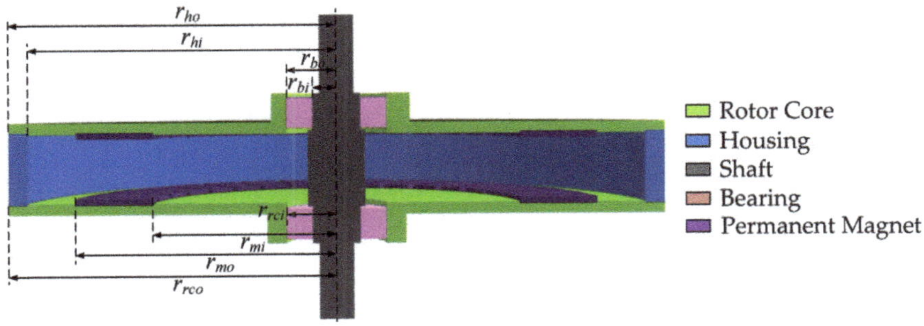

Figure 12. Cross-section of the YASA machine.

3.2.2. Bearing

The bearing connects the stationary part and the rotating part. Figure 13 shows the thermal model of the bearing. P_b represents the friction loss of the bearing. A part of the heat transfers from the bearing to the rotor core and is dissipated. R_{shb} is the thermal resistance between the bearing and shaft, as shown in Figure 9. R_{brc} is the thermal resistance between the bearing and rotor core, as shown in Figure 13. The equations follow:

$$R_{shb} = \frac{(r_{bo} - r_{bi})/2}{\frac{\left(\frac{r_{bo}+r_{bi}}{2}+r_{bi}\right)}{2}2\pi H_{be}k_{be}} = \frac{(r_{bo} - r_{bi})/2}{\left(\frac{r_{bo}+r_{bi}}{2} + r_{bi}\right)\pi H_{be}k_{be}} \tag{24}$$

$$R_{brc} = \frac{(r_{bo} - r_{bi})/2}{\left(\frac{r_{bo}+r_{bi}}{2} + r_{bo}\right)\pi H_{be}k_{be}} \tag{25}$$

where r_{bo} and r_{bi} are the outer and inner radius of the bearing, respectively, and k_{be} is the equivalent thermal conductivity of the bearing.

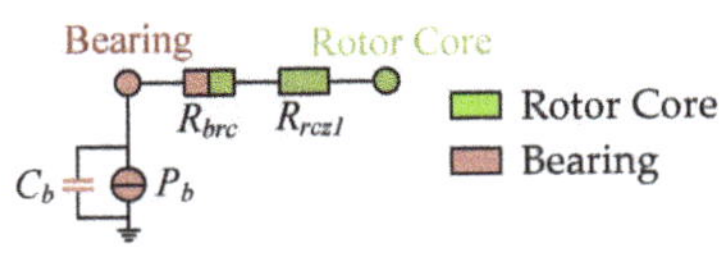

Figure 13. Bearing thermal model.

R_{rcz1} is the thermal resistance of the rotor core. Assuming the rotor core and the permanent magnet are the same distance to the shaft, then:

$$R_{rcz1} = \frac{r_{rcma} - r_{rci}}{(r_{rcma} + r_{rci})\pi H_{rc}k_{rc}} \tag{26}$$

where r_{rcma} is the distance from the permanent magnet to the shaft center, R_{rci} is the inner radius of the rotor core, which is equal to the outer radius of the bearing, H_{rc} is the height of the rotor core, and k_{rc} is the thermal conductivity of the rotor core.

3.2.3. Rotor Core

The eddy current loss of the rotor core is calculated by the FE method. Figure 14 shows the thermal model of the rotor core, which contacts the inner-air and surroundings directly. R_{rcia} is the convective thermal resistance between the rotor core and inner-air, R_{rcso} is the convective thermal resistance between the rotor core and surroundings, R_{rcz2} is the thermal resistance of the rotor core, and R_{hy} is the axial thermal resistance of the housing. The calculation formulae follow:

$$R_{rcia} = \frac{1}{k_{rcia}S_{rcia}} \tag{27}$$

$$S_{rcia} = \pi\left(r_{hi}^2 - r_{mo}^2\right) + \pi\left(r_{mi}^2 - r_{rci}^2\right) \tag{28}$$

$$R_{rcso} = \frac{1}{k_{rcso}S_{rcso}} \tag{29}$$

$$S_{rcso} = \pi\left(r_{rco}^2 - r_{rci}^2\right) \tag{30}$$

$$R_{rcz2} = \frac{r_{rco} - r_{rcma}}{(r_{rcma} + r_{rco})\pi H_{rc}k_{rc}} \tag{31}$$

$$R_{hy} = \frac{H_h/4}{\pi(r_{ho}^2 - r_{hi}^2)k_h} \tag{32}$$

where k_{rci} is the convective heat transfer c between the rotor core and inner-air, and k_{rcso} is the convective efficient heat transfer coefficient between the rotor core and surroundings. S_{rcia} is the contact area between the rotor core and inner-air, S_{rcso} is the contact area between the rotor core and surroundings; r_{mo} and r_{mi} are the outer and inner radius of the permanent magnet, respectively; r_{ho} and r_{hi} are the outer and inner radius of the housing, respectively; H_h is the height of the housing; and k_h is the thermal conductivity of the housing.

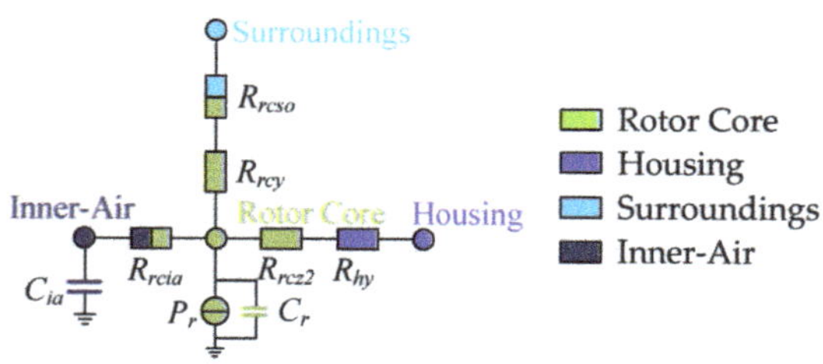

Figure 14. Thermal model of the rotor core.

3.2.4. Housing

There is heat convection between the inner surface of the housing and inner-air. The outer surface contacts with the surroundings, which dissipate heat directly. Figure 15 shows the thermal model of the housing. R_{hz} is the radial thermal resistance of housing, R_{hia} is the convective thermal resistance between the housing and inner-air, and R_{hso} is the convective thermal resistance between the housing and surroundings. The formulas follow:

$$R_{hia} = \frac{1}{S_{hia}k_{hia}} \tag{33}$$

$$S_{hia} = \pi r_{hi}H_h \tag{34}$$

$$R_{hso} = \frac{1}{S_{hso}k_{hso}} \tag{35}$$

$$S_{hso} = \pi r_{ho}H_h \tag{36}$$

$$R_{hz} = \frac{(r_{ho} - r_{hi})/2}{\left(\frac{r_{ho}+r_{hi}}{2}\right)2\pi \frac{H_h}{2} k_h} = \frac{r_{ho} - r_{hi}}{(r_{ho} + r_{hi})\pi H_h k_h} \tag{37}$$

where k_{hia} is the convective heat transfer coefficient between the housing and inner-air, k_{hso} is the convective heat transfer coefficient between the housing and surroundings, S_{hia} is the contact area between the housing and inner-air, and S_{hso} is the contact area between the housing and surroundings.

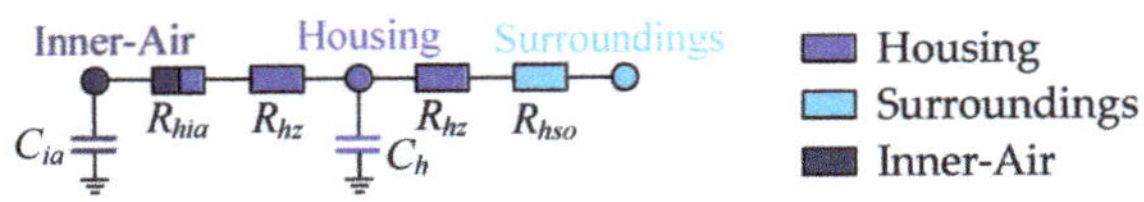

Figure 15. Housing thermal model.

3.3. Entire Thermal Model of the YASA Machine

Based on the thermal models of all the components described above, the entire thermal model of the YASA machine can be obtained by simply integrating all of them, as shown in Figure 16. The thermal model of each component marked with different colors is connected by thermal resistances (including conduction thermal resistance and convective thermal resistance). Between the shaft and the supporting frame, there is the conduction thermal resistance R_{sush}. The thermal model of the bearing and shaft is connected by the thermal resistance R_{shb}. The permanent magnet adheres to the rotor core, so there is the conduction thermal resistance R_{mrc}. Additionally, there are the convective thermal resistances (including R_{mia}, R_{scia}/Q, and R_{wia}/Q) of the YASA machine between the permanent magnet, stator core, armature winding, and inner-air. These thermal resistances are calculated by the formulae given above.

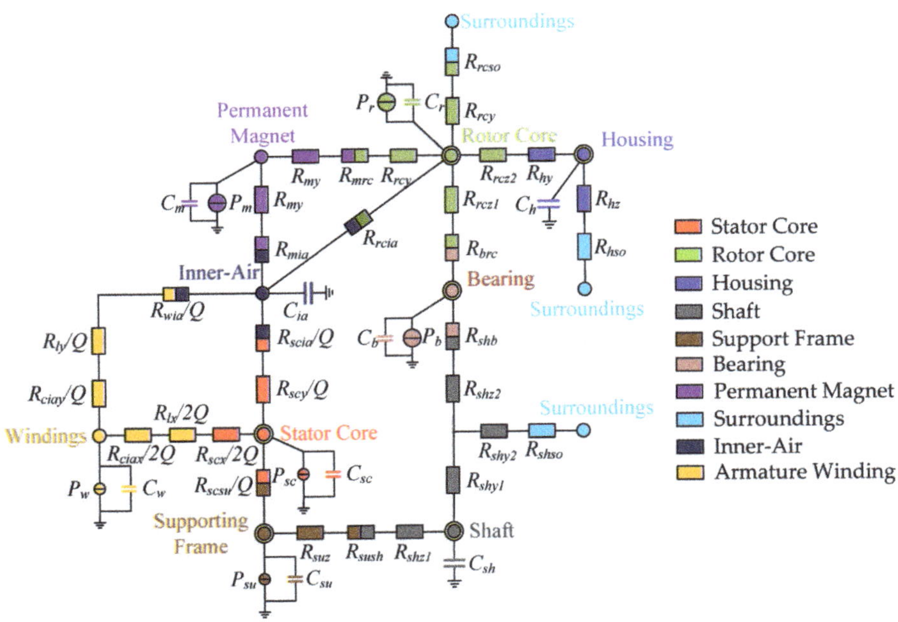

Figure 16. Entire thermal model of the YASA machine.

4. Parameter Calculation of the Thermal Network Model

In this section, the losses of the YASA machine (including the armature winding loss, stator core loss, rotor core loss, permanent magnet eddy current loss, and mechanical loss), the thermal resistance of the air gap, and convective heat transfer coefficient on the hub surface are calculated. Finally, the detailed temperature distribution of the YASA machine is obtained.

4.1. Losses Calculation

The copper loss is calculated by

$$P_{copper} = mI_s^2R_s \tag{38}$$

where m is the phase number, R_s is the phase resistance, and I_s is the effective value of phase current.

The core loss separation model proposed by Italian scholar Bertotti [26] is widely used. It mainly includes hysteresis loss, classical eddy current loss, and excess loss:

$$P_{Fe} = P_h + P_c + P_e = K_h f B_m^{\propto} + \frac{\sigma d^2}{12\rho}\frac{1}{T}\int_0^T\left(\frac{dB(t)}{dt}\right)^2 dt + \frac{\sqrt{\sigma G V_0 S}}{\rho}\frac{1}{T}\int_0^T\left(\frac{dB(t)}{dt}\right)^{1.5} dt \tag{39}$$

The core loss is related to magnetic flux density. When the magnetic flux density changes sinusoidally and remagnetization is adopted, Equation (39) can be simplified as

$$P_{Fe} = K_h f B_m^{\propto} + K_c f^2 B_m^2 + K_e f^{1.5} B_m^{1.5} \tag{40}$$

where K_h, K_c, and K_e are the coefficient of hysteresis loss, eddy current loss, and excess loss, respectively; f is the electrical frequency and B_m is the amplitude of magnetic flux density.

Eddy current loss of the rotor core and permanent magnet can be calculated by

$$P_{eddy} = \frac{1}{\sigma}\int J^2 dV \tag{41}$$

where σ is the conductivity, J is the eddy current density, and V is the volume of the component.

Mechanical loss includes bearing loss P_b and wind friction loss P_{wind}. The formula follows:

$$P_b = 0.06k_{fb}(m_r + m_{sh})n_s \tag{42}$$

where k_{fb} is an empirical coefficient, which in the range 1–3 m^2/s^2; m_r, m_{sh} is the mass of rotor and shaft, respectively; n_s is the rotating speed.

$$P_{wind} = \frac{1}{2}c_f\rho_{cm}(2\pi n_s)^3\left(R_{out}^5 - R_{sh}^5\right) \tag{43}$$

where ω_s is electric angular speed; V_{air} is the dynamic viscosity of air, taken as 2×10^{-5} Pas; and ρ_{cm} is the air density, taken as 1.2 kg/m^3. R_{out} is the outer radius of the rotor and R_{sh} is the outer radius of the shaft.

4.2. Air Gap Thermal Resistance and Convective Heat Transfer Coefficient on the Hub Surface

The calculation process of the air gap thermal resistance follows four points:

(1) Reynolds number R_{eg} of the air gap is determined according to machine radius, speed, and dynamic viscosity of the fluid, as shown in Equation (44) [27].

(2) The fluid types are determined by Reynolds number R_{eg}, which generally includes laminar flow, transition flow, and turbulent flow [28].

(3) Determination of the Nusselt constant N_u is based on Reynolds number R_{eg} and the air gap ratio G ($G = g/R_2$, where g is the distance between the stator and rotor plate, R_2 is the outer radius of the rotating plate), as shown in Table 2 [29].

Table 2. Nusselt constant function relationship.

Value Range	Fluid Type	Formula
G = 0.01	Laminar flow	$Nu = 7.46Re_g^{0.32}$
$0.02 \leq G \leq 0.06$	Laminar flow	$Nu = 0.50(1 + 5.47 \times 10^{-4}exp(112G))Re_g^{0.5}$
$G \geq 0.06$	Laminar flow	$Nu = 0.55\left(1 + 0.462exp\left(\frac{-13G}{3}\right)\right)Re_g^{0.5}$
Rotor plate	Laminar flow	$Nu = 0.55Re_g^{0.5}$
G = 0.01	Turbulence	$Nu = 0.044Re_g^{0.75}$
$0.02 \leq G \leq 0.06$	Turbulence	$Nu = 0.033(12.57exp(-33.18G))Re_g^{3/5+25G^{12/7}}$
$G \geq 0.06$	Turbulence	$Nu = 0.0208(1 + 0.298exp(-9.27G))Re_g^{0.8}$
Rotor plate	Turbulence	$Nu = 0.0208Re_g^{0.8}$

(4) According to this derivation, the air-gap equivalent convective heat transfer coefficient h_g and the air-gap thermal resistance R_g can be obtained by Formulae (46) and (47).

$$Re_g = \frac{\omega_s R_{out}^2}{v_{air}} \tag{44}$$

$$c_f = \frac{3.87}{\sqrt{Re_g}} \tag{45}$$

$$h_g = \frac{Nu \times k_{air}}{R_2} \tag{46}$$

$$R_g = \frac{1}{\pi\left(R_2^2 - R_1^2\right)h_g} \tag{47}$$

where k_{air} is the thermal conductivity of air and R_1 is the inner radius of the rotating plate.

There is no barrier between the end cap and the surroundings, which belongs to heat convection. The convective heat transfer coefficient on the surface of the rotor plate can be obtained by Equation (46).

The average Nusselt constant is obtained from two cases. One is that the airflow type on the surface of the rotor plate is laminar flow, so the average Nusselt constant N_{u1} is calculated by

$$Nu_1 = \frac{2}{5}\left(Re_g^2 + Gr\right)^{\frac{1}{4}} \tag{48}$$

$$Gr = \frac{\beta g R^3 \pi^{3/2} \Delta T}{v^2} \tag{49}$$

where β is the coefficient of thermal expansion; v is the kinematic viscosity coefficient of the fluid, taken as 1.569×10^{-5} m^2/s. ΔT is the temperature difference between the rotor plate surface and the surroundings.

The other airflow type is a combination of laminar flow and turbulence:

$$Nu_1 = 0.15Re_g^{\frac{4}{5}} - 100\left(\frac{r_c}{R_2}\right)^2 \tag{50}$$

$$r_c = \sqrt{2.5 \times 10^5 v/\omega_s} \tag{51}$$

where r_c is the transition radius.

In addition, with the help of the naphthalene sublimation experiment and analogy theory, the relation between the convective heat transfer coefficient and the speed of the rotor plate is obtained by fitting. This paper refers to this method to obtain the convective heat transfer coefficient of the end cap:

$$h_g = 8.859\omega_s^{0.5} \tag{52}$$

The convective heat transfer between the housing and surroundings is calculated by

$$h_p = \frac{k_{air}}{D_2} Nu_2 \quad (53)$$

$$Nu_2 = 0.133 Re_D^{2/3} Pr^{1/3} \quad (54)$$

$$Re_D = \frac{\omega_s D_2^2}{v} \quad (55)$$

where h_p, $N_{u2,}$ and Re_D are the convective heat transfer coefficient, average Nusselt constant, and Reynolds number of the rotor plate, respectively. D_2 is the outer diameter of the rotor plate; P_r is the Prandtl number of the air, taken as 0.703.

The inner-air thermal resistance, the convective heat transfer coefficient between the rotor core and surroundings, and the convective heat transfer coefficient between the housing and surroundings are 0.7568 K/w, 62.8 w/(m^2·K), and 47.8 w/(m^2·K), respectively.

4.3. Results of the Thermal Network Model

The temperature distribution of the YASA machine can be obtained by the entire thermal network model. The temperature of all the components is listed in Table 3.

Table 3. The steady-state temperature of the YASA machine.

Machine Components	Steady-State Temperature (°C)
Winding	127.07
Stator core	100.57
Permanent magnet	54.54
Rotor core	52.00
Housing	49.35
Bearing	63.61

It can be found that the armature winding temperature is the highest and the temperature of the stator core is lower than that of the winding temperature. The lowest temperature is the housing and rotor core. The armature winding is the main heat source of the YASA machine. The stator core and winding are located in the center of the machine, from which dissipation of heat is difficult. The rotor core and the housing are contacted directly by the surroundings, which makes its lower temperature rise. Under the rated condition, the temperature of the permanent magnet is 54.54 °C. The temperature does not exceed the allowable working temperature, which ensures reliable operation.

By setting the laboratory temperature at 20 °C, all components of the YASA machine tend to be stable at 80 min, as shown in Figure 17.

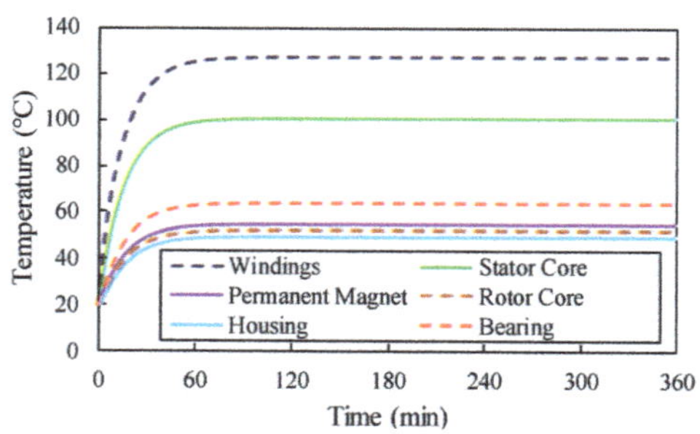

Figure 17. The transient temperature rise of the YASA machine.

5. Experimental Verification

5.1. Experimental Prototype

The prototype is shown in Figure 18. The segmented stator core and armature winding are fixed by the support frame, as shown in Figure 18a. Permanent magnets are bonded to the rotor core, as shown in Figure 18b. The YASA machine assembled is shown in Figure 18c.

(**a**) (**b**) (**c**)

Figure 18. YASA machine assembly: (**a**) stator; (**b**) rotor; (**c**) machine assembly.

5.2. Prototype Experiment

The rated 11 A direct current (DC) corresponding 5 A/mm^2 current density is applied to the armature winding. The experimental data of the YASA machine are compared with the calculated results of the LPTN model, and the results are in good agreement, as shown in Figure 19.

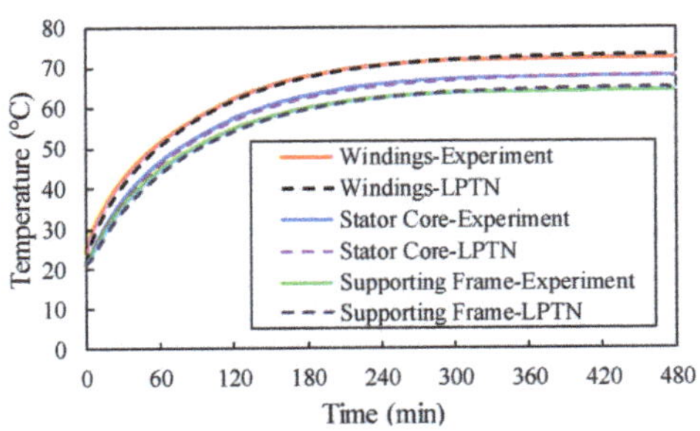

Figure 19. Instantaneous temperature-rise test of the prototype under 11 A DC.

Further, the direct current with different values is applied to the windings to measure the temperature rise of the YASA machine. The current is disconnected after a period of time, the cooling curve is measured, and the results are compared with the thermal network, as shown in Figures 20–22. The maximum error is less than 2.5 °C, which occurs in the armature windings. The error mainly comes from the change in environmental temperature and the given error of heat capacity. The error is smaller than that of the LPTN model for the single-sided AFPM machine proposed in [17], whose temperature error is 4 °C.

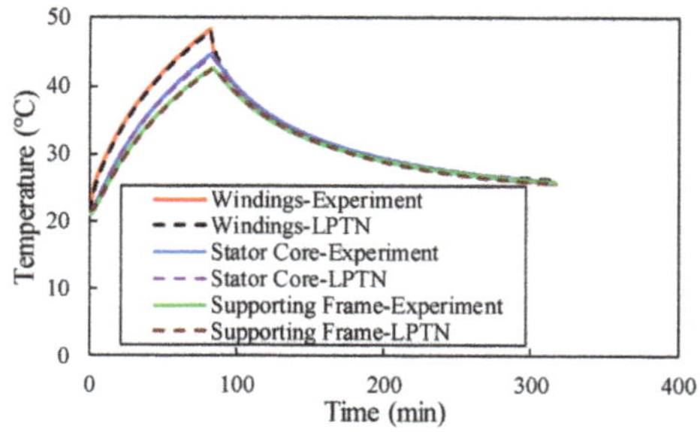

Figure 20. Instantaneous temperature-rise test of the prototype under 10 A DC.

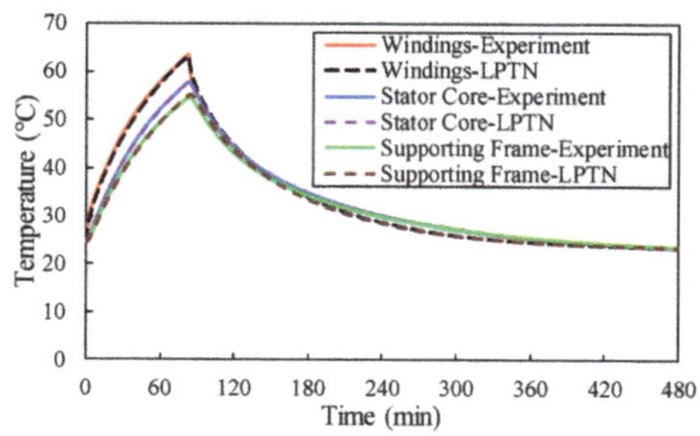

Figure 21. Instantaneous temperature-rise test of the prototype under 12 A DC.

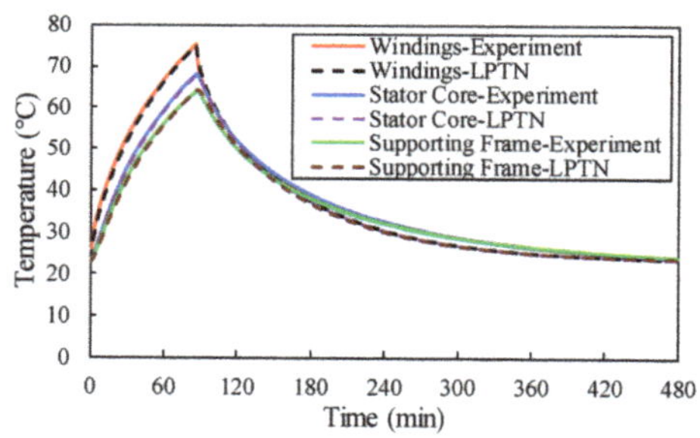

Figure 22. Instantaneous temperature-rise test of the prototype under 13.5 A DC.

6. Summary

In this paper, the LPTN model of the YASA machine for in-wheel traction application is developed to calculate the temperature distribution of all components of the YASA machine. The thermal models of all components in YASA machines, including the stationary and rotary components, are simplified based on the symmetrical structure, while the detailed thermal resistance formulae are given. Based on the loss results calculated by electromagnetic FE analysis, the temperature distribution of the YASA machine is obtained. Comparing the calculated and experimental results, the maximum temperature difference is no more than 3.3%, which validates very good accuracy of the proposed thermal model. The proposed method is considered as a good reference for design engineers of YASA machines in the applications of in-wheel traction. In addition, this paper is also beneficial to the research of machine cooling. The advanced cooling technique of the YASA machine applied on in-wheel traction systems will be investigated in a following study.

Author Contributions: Conceptualization, H.Z.; methodology, G.W.; validation, Y.W. and G.W.; investigation, G.W., Y.W. and Q.N.; data curation, Y.W.; writing—original draft preparation, G.W.; writing—review and editing, Y.G. and Q.N.; visualization, W.H.; supervision, W.H. and H.Z.; project administration, W.H. All authors have read and agreed to the published version of the manuscript.

Funding: This research was funded by The National Key Research and Development Program of China grant number 2021YFB2500703.

Institutional Review Board Statement: Not applicable.

Informed Consent Statement: Not applicable.

Data Availability Statement: Not applicable.

Conflicts of Interest: The authors declare no conflict of interest.

References

1. Chang, J.; Fan, Y.; Wu, J. A Yokeless and Segmented Armature Axial Flux Machine with Novel Cooling System for In-Wheel Traction Applications. *IEEE Trans. Ind. Electron.* **2021**, *68*, 4131–4140. [CrossRef]
2. Taran, N.; Klink, D.; Heins, G. A Comparative Study of Yokeless and Segmented Armature Versus Single Side Axial Flux PM Machine Topologies for Electric Traction. *IEEE Trans. Ind. Appl.* **2022**, *58*, 325–335. [CrossRef]
3. Li, T.; Liang, X.; Dou, H. Thermal Analysis of the Yokeless and Segmented Armature Axial Flux In-wheel motor. In Proceedings of the 2020 International Conference on Artificial Intelligence and Computer Engineering (ICAICE), Beijing, China, 23–25 October 2020; pp. 449–452.
4. Zhang, B.; Seidler, T.; Dierken, R.; Doppelbauer, M. Development of a Yokeless and Segmented Armature Axial Flux Machine. *IEEE Trans. Ind. Electron.* **2016**, *63*, 2062–2071. [CrossRef]
5. Xu, L.; Xu, Y.; Gong, J. Analysis and Optimization of Cogging Torque in Yokeless and Segmented Armature Axial-Flux Permanent-Magnet Machine with Soft Magnetic Composite Core. *IEEE Trans. Magn.* **2018**, *54*, 8106005. [CrossRef]
6. Geng, W.; Zhang, Z.; Li, Q. High Torque Density Fractional-Slot Concentrated-Winding Axial-Flux Permanent-Magnet Machine with Modular SMC Stator. *IEEE Trans. Ind. Appl.* **2019**, *56*, 3691–3699.
7. Geng, W.; Wang, Y.; Wang, J. Comparative Study of Yokeless Stator Axial-Flux PM Machines having Fractional Slot Concentrated and Integral Slot Distributed Windings for Electric Vehicle Traction Applications. *IEEE Trans. Ind. Electron.* 2022; *early access*.
8. Wang, Y.; Lu, J.; Liu, C.; Lei, G.; Guo, Y.; Zhu, J. Development of a High-Performance Axial Flux PM Machine with SMC Cores for Electric Vehicle Application. *IEEE Trans. Magn.* **2019**, *55*, 8105304. [CrossRef]
9. Kim, J.; Lee, J.; Song, J. Characteristics Analysis Method of Axial Flux Permanent Magnet Motor Based on 2-D Finite Element Analysis. *IEEE Trans. Magn.* **2017**, *53*, 8105304. [CrossRef]
10. Gerlando, A.; Foglia, G.; Iacchetti, M.; Perini, R. Parasitic Currents in Stray Paths of Some Topologies of YASA AFPM Machines: Trend with machine size. *IEEE Trans. Ind. Electron.* **2016**, *63*, 2746–2756. [CrossRef]
11. Camilleri, R.; Howey, D.A.; McCulloch, M.D. Predicting the Temperature and Flow Distribution in a Direct Oil-Cooled Electrical Machine with Segmented Stator. *IEEE Trans. Ind. Electron.* **2016**, *63*, 82–91. [CrossRef]
12. Vansompel, H.; Leijnen, P.; Sergeant, P. Multiphysics Analysis of a Stator Construction Method in Yokeless and Segmented Armature Axial Flux PM Machines. *IEEE Trans. Energy Convers.* **2019**, *34*, 139–146. [CrossRef]
13. Fawzal, A.S.; Cirstea, R.M.; Gyftakis, K.N. Fan Performance Analysis for Rotor Cooling of Axial Flux Permanent Magnet Machine. *IEEE Trans. Ind. Appl.* **2017**, *53*, 3295–3304. [CrossRef]
14. Luo, L.; Chang, J.; Wu, J. Design and Analysis of a Water-Cooling System in a New Yokeless and Segmented Armature Axial In-Wheel Motor for Electric Vehicles. *J. Therm. Sci. Eng. Appl.* **2021**, *13*, 1–12. [CrossRef]
15. Le, W.; Lin, M.; Lin, K. A Novel Stator Cooling Structure for Yokeless and Segmented Armature Axial Flux Machine with Heat Pipe. *Energies* **2021**, *14*, 5717. [CrossRef]
16. Le, W.; Lin, M.; Jia, L. Design of a Novel Stator Water-Cooling System for Yokeless and Segmented Armature Axial Flux Machine. In Proceedings of the 2021 IEEE 4th Student Conference on Electric Machines and Systems (SCEMS), Huzhou, China, 2–3 December 2021; pp. 1–4.
17. Winterborne, D.; Stannard, N.; Sjöberg, L. An Air-Cooled YASA Motor for in-Wheel Electric Vehicle Applications. *IEEE Trans. Ind. Appl.* **2020**, *56*, 6448–6455. [CrossRef]
18. Vansompel, H.; Hemeida, A.; Sergeant, P. Stator Heat Extraction System for Axial Flux Yokeless and Segmented Armature Machines. In Proceedings of the 2017 IEEE International Electric Machines and Drives Conference (IEMDC), Miami, FL, USA, 21–24 May 2017; pp. 1–7.
19. Burke, R.; Giedymin, A.; Wu, Z. A Lumped Parameter Thermal Model for Single-Sided AFPM Machines with Experimental Validation. *IEEE Trans. Ind. Electron.* **2020**, *6*, 1065–1083. [CrossRef]
20. Mohamed, A.; Hemeida, A.; Vansompel, H. Parametric Studies for Combined Convective and Conductive Heat Transfer for YASA Axial Flux Permanent Magnet Synchronous Machines. *Energies* **2018**, *11*, 2983. [CrossRef]

21. Mohamed, A.H.R.; Vansompel, H.; Hemeida, A. Transient Lumped Parameter Thermal Model Based Load ability Studies for The YASA Axial Flux Permanent Magnet Synchronous Machines. In Proceedings of the Young Researchers Symposium, Brussels, Belgium, 25–25 May 2018.
22. Chen, Q.; Liang, D.; Gao, L. Hierarchical Thermal Network Analysis of Axial-flux Permanent-Magnet Synchronous Machine for Electric Motorcycle. *IET Electr. Power Appl.* **2018**, *12*, 859–866. [CrossRef]
23. Mohamed, A.H.; Hemeida, A.; Rashekh, A. A 3D Dynamic Lumped Parameter Thermal Network of Air-Cooled YASA Axial Flux Permanent Magnet Synchronous Machine. *Energies* **2018**, *11*, 774. [CrossRef]
24. Zhang, H.; Giangrande, P.; Sala, G. Thermal Model Approach to Multisector Three-Phase Electrical Machines. *IEEE Trans. Ind. Electron.* **2021**, *4*, 2919–2930. [CrossRef]
25. Staton, D.; Boglietti, A.; Cavagnino, A. Solving the More Difficult Aspects of Electric Motor Thermal Analysis in Small and Medium Size Industrial Induction Motors. *IEEE Trans. Energy Convers.* **2005**, *20*, 620–628. [CrossRef]
26. Bertotti, G. General properties of power losses in soft ferromagnetic materials. *IEEE Trans. Magn.* **1988**, *24*, 621–630. [CrossRef]
27. Howey, D.A.; Childs, P.R.N.; Holmes, A.S. Air-gap Convection in Rotating Electrical Machines. *IEEE Trans. Ind. Electron.* **2012**, *59*, 1367–1375. [CrossRef]
28. Owen, J.M.; Haynes, C.M.; Bayley, F.J. Heat Transfer from an Air-Cooled Rotating Disk. *Proc. R. Soc. Lond. A Math. Phys. Sci.* **1974**, *336*, 453–473.
29. Boutarfa, R.; Harmand, S. Local Convective Heat Transfer for Laminar and Turbulent Flow in a Rotor-Stator System. *Exp. Fluids* **2005**, *38*, 209–221. [CrossRef]

energies

Article

Optimal Control of the Diesel–Electric Propulsion in a Ship with PMSM

Zenon Zwierzewicz [1], Dariusz Tarnapowicz [1,*], Sergey German-Galkin [1] and Marek Jaskiewicz [2]

[1] Faculty of Mechatronics and Electrical Engineering, Maritime University of Szczecin, 70-500 Szczecin, Poland
[2] Faculty of Mechatronics and Mechanical Engineering, Kielce University of Technology, 25-314 Kielce, Poland
* Correspondence: d.tarnapowicz@pm.szczecin.pl; Tel.: +48-9148-09-955

Abstract: The article presents and compares two different control methods for a permanent magnet synchronous motor (PMSM) for diesel–electric ship propulsion. The main focus of the article is on control optimization, which allows improving energy efficiency by reducing reactive power in the mechatronic propulsion system. The first method consists in modifying the commonly used field-oriented control (FOC) strategy to ensure zero reactive power in the inverter–PMSM system. Since a characteristic of ship propulsion systems, unlike those used on land, is the step load on the propulsion motor, the system's performance in dynamic states is particularly important. Unfortunately, control strategies based on FOC do not take into account the dynamics of the system, since they apply only to steady states. Therefore, the authors of this paper, based on control theory methods, proposed an approach that also optimizes control in dynamic states, while minimizing reactive power in the steady state. The analytical studies were confirmed in simulation studies using the MATLAB Simulink package.

Keywords: diesel–electric propulsion; permanent magnet synchronous motor; energy optimization; I/O linearization method

Citation: Zwierzewicz, Z.; Tarnapowicz, D.; German-Galkin, S.; Jaskiewicz, M. Optimal Control of the Diesel–Electric Propulsion in a Ship with PMSM. *Energies* **2022**, *15*, 9390. https://doi.org/10.3390/en15249390

Academic Editors: Gan Zhang and Hao Hua

Received: 10 November 2022
Accepted: 9 December 2022
Published: 12 December 2022

1. Introduction

Due to the fact that approximately 80% of world trade is carried out by sea [1], maritime organizations and designers of ship systems are looking for solutions limiting the negative effects of ships on the environment. In this context, the International Maritime Organization (IMO) has presented rigorous regulations concerning the design of ship systems by introducing the Energy Efficiency Design Index (EEDI) [2–5]. The introduced regulations pose a great challenge for engineers, who (at every stage of ship design) are looking for solutions to improve the EEDI by applying new technologies or increasing the efficiency of currently applied systems.

The most frequently used ship propulsion system is an internal combustion engine mechanically connected to a propeller with a constant pitch or with a variable pitch (less frequently). The main disadvantage of internal combustion engines is their low efficiency—especially at low engine loads. The typically used low-speed diesel engines show up to 50% efficiency at optimal load [6]. It should be stressed that ships with combustion engines and fixed propellers have maneuvering limitations.

Nowadays, this propulsion is a solution commonly used for ships, which move at design speed [7] for most of their operation.

For ships that require high flexibility of the propulsion system, e.g., cruise ships, ferries, pleasure vessels, and offshore ships (anchor handling tug vessels, accommodation ships, supply vessels), thanks to the development of power electronics technologies, diesel–electric propulsion systems have been an alternative solution to internal combustion engines. In addition, this drive is characterized, among other things, by high reliability and overall cost reduction (investment and exploitation) [7,8].

The traditionally used power generation system and the ship propulsion in the diesel–electric system are presented in Figure 1 [7,9–13].

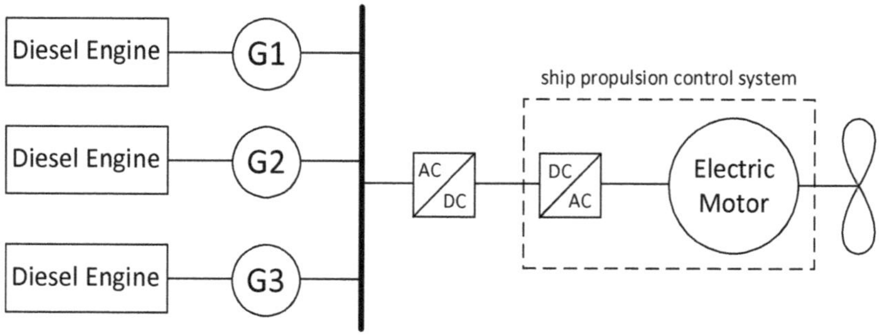

Figure 1. Traditionally used ship propulsion system in the diesel–electric system (AC grid).

Compared to the propulsion of a ship with an internal combustion engine, the efficiency of the diesel–electric propulsion at full load is lower due to additional losses of the power electronics system and the electric motor (approximately 4% [7,14]). However, with a reduced load, the efficiency of the diesel–electric system is greater. This is due to the operation of a smaller number of generating sets, and thus the optimal operation of the remaining generating sets. In ships with AC grids, all generating units operate at a constant speed in order to ensure the stability and safety of the ship's power grid.

Increasing the efficiency of the diesel–electric system and therefore increasing the efficiency of the drive can be achieved by using the topology shown in Figure 2.

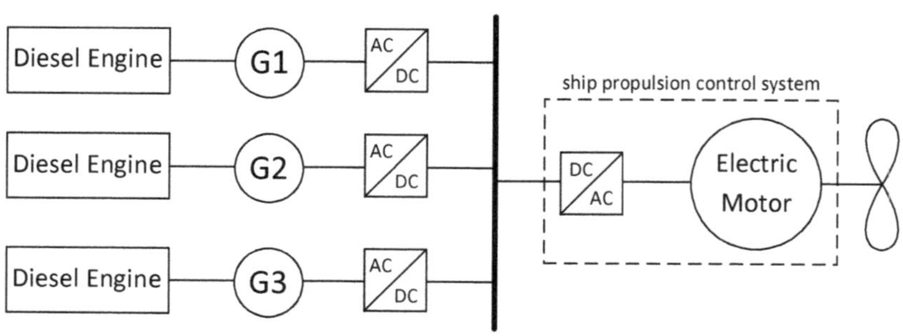

Figure 2. Ship propulsion system in the diesel–electric system with optimization of the operation of generator propulsion engines (DC grid).

The ship propulsion system with optimized operation of generator propulsion engines is also known (Figure 2) [7,15–20].

In the DC grid, it is possible to optimize the operation of generator propulsion engines. Each diesel engine can run independently in the DC grid at optimal speed. Therefore, compared to the traditional diesel–electric propulsion system with AC grid technology (Figure 1), the ship's propulsion system with DC grid technology (Figure 2) shows greater efficiency.

The use of different types of electric motors in the diesel–electric system is also associated with obtaining different levels of efficiency of the propulsion system.

The most commonly used motor in the ship's electric propulsion is the externally excited synchronous motor (EESM). Asynchronous AC motors and DC motors, despite attempts to use them in the electric propulsion of ships, were not widely used due to technological (design) limitations—especially for very high-power propulsion systems. The continuous development in the technology of producing permanent magnets and the huge progress in power electronics technology made it possible to use PMSM in the electric propulsion of the ship. In addition to the well-known advantages of using PMSM motors (e.g., higher power density, high reliability, lower weight and dimensions), these motors are mainly characterized by higher efficiency (especially when the motor is underloaded) due mainly to the elimination of excitation losses (from 3 to 6%, depending on the load). Thanks to the possibility of using a large number of pairs of poles, the lack of a need for gears (gear losses of about 2%, depending on the number of gears used) also increases efficiency [21,22].

Currently, the main problem for designers of ship systems is the increase in the energy efficiency of systems. Engineers are faced with the difficult task of limiting the ship's energy efficiency design index (EEDI), which is stipulated by regulations. It is very difficult to achieve.

However, the presented above-selected methods of increasing the efficiency of diesel–electric propulsion of the ship, and thus increasing energy efficiency, do not completely solve the problem.

There are known methods of improving the energy efficiency of control of a PMSM propulsion system by expanding the topology of the power electronic system [23] or complex control algorithms [24–26] in which the response of the system to abrupt (rapid) large changes in the load and speed of the motor, which occur on the ship, is omitted.

Therefore the authors in this article propose two new methods of controlling the PMSM motor in the diesel–electric propulsion system of a ship, increasing the propulsion efficiency by reducing the reactive power in the inverter–PMSM system. The first method presented in the article is based on steady-state analysis, the effectiveness of which was confirmed by comparison with the classical FOC method of PMSM control. The second method of control is based on the output feedback linearization [27] and LQR optimal control [28] methods. This approach, in addition to reducing reactive power in the steady state, also makes it possible to reduce torque oscillations in dynamic states (which is the case with the first method presented).

The paper is organized as follows:

Section 2.1 presents the steady-state PMSM control method based on the use of the physical properties of the system with energy optimization and presents the results of the study.

Section 2.2 discusses the details of the analysis of the I/O linearization method providing energy optimization and further ensuring optimal steady-state behavior of the system.

The article concludes in Section 3 with comments on the results.

Section 4 summarized the effectiveness of the two methods in terms of energy efficiency and the second method of optimization in the dynamic state.

2. Methods and Results

2.1. Energy Optimization of the PMSM Propulsion System in a Steady State Based on the Use of the Physical Properties of the System

This subsection will present an energy optimization method based on the analysis of steady-state electromagnetic processes, which was compared to the classic FOC control method. Analysis and simulation studies confirming the effectiveness of the method were carried out using the MATLAB Simulink program.

The equivalent electrical diagram of the PMSM propulsion system is presented in Figure 3.

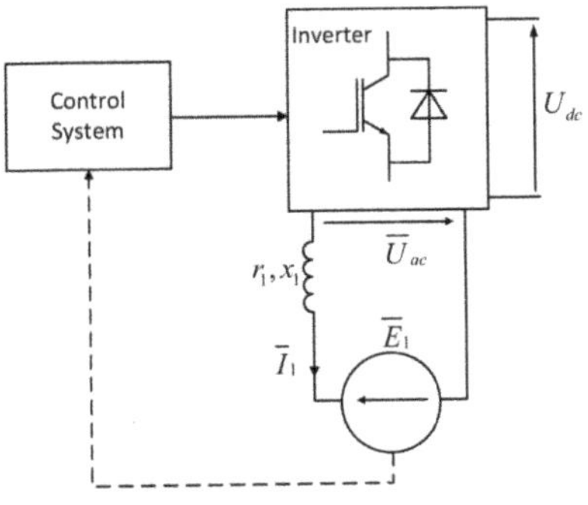

Figure 3. Equivalent diagram of the propulsion system with PMSM.

The PMSM control system is based on the selection of a coordination system in such a way that variables defining its dynamics are constant in a steady state.

The UM mathematical description is conducted in the synchronously rotating system of d, q coordinates, which are closely connected with the PMSM magnetic field. The d axis coincides with the magnetic field vector. Then the electromagnetic force E will coincide with the q axis. The nonlinear model of the PMSM machine can be written in the form of a system of differential Equation (1).

$$\begin{gathered} U_{acd}(t) = r_1 I_d(t) + L_1 \tfrac{dI_d(t)}{dt} - \omega L_1 I_q(t), \\ U_{acq}(t) = r_1 I_q(t) + L_1 \tfrac{dI_q(t)}{dt} + \omega L_1 I_d(t) + \psi_0 \omega, \\ T_e = 1.5 p \psi_0 I_q(t), \\ J \tfrac{d\omega_m}{dt} = T_e - T_L. \end{gathered} \tag{1}$$

where:

$U_{acd}(t)$, $U_{acq}(t)$, $I_d(t)$, $I_q(t)$—the inverter output voltage and stator currents (PMSM) on the d and q axes;

$\omega = p\omega_m$—voltage pulsation at the inverter output;

ω_m—angular speed of the machine's rotor;

J—inertia;

p—number of machine's pole pairs;

T_e—electromagnetic moment of the machine;

T_L—load torque;

r_1—stator winding phase resistance;

$x_1 = \omega L_1 = p\,\omega_m L_1$—inductive reactance of the stator winding phase (L_1—inductance of the stator winding phase).

By analyzing the system (1), it is easy to notice that for specified ω_m and T_L there are an infinite number of equilibrium points (steady states) of the system, associated with input pairs (U_{acd}, U_{acq}).

Therefore, there is a natural problem of selecting a pair of inputs (controls) $U_{acd}{}^*$ and $U_{acq}{}^*$ in such a way as to ensure the reset of the system's reactive power ($Q_{ac} = 0$), which we will refer to as energy optimization [21,29,30].

For this purpose, we conduct the following analysis:

In the steady state, electromagnetic processes presented by Equation (1), describing the propulsion, take the following form:

$$\begin{gathered} U_{acd} = r_1 I_d - x_1 I_q, \\ U_{acq} = r_1 I_q + x_1 I_d + E_q. \end{gathered} \tag{2}$$

where:

$E_q = p\omega_m \psi_0 = \omega \psi_0$—electromotive force of the PMSM stator.

The Equation (2) can be presented in a vector form as follows:

$$\overline{U}_{ac} = \overline{E} + r_1 \overline{I}_1 + j x_1 \overline{I}_1 \tag{3}$$

where:

$\overline{U}_{ac}$, $\overline{I}_1$, $\overline{\psi}_0$, $\overline{E} = j\omega\overline{\psi}_0$—resultant (spatial) vector of variable electromagnetic states of the system.

In the mathematical description, the variables in Equation (3) are vector values. This enables building a vector diagram and evaluating the physical properties of the system. Figure 4 shows the vector diagram for the inverter–PMSM system with energy optimization.

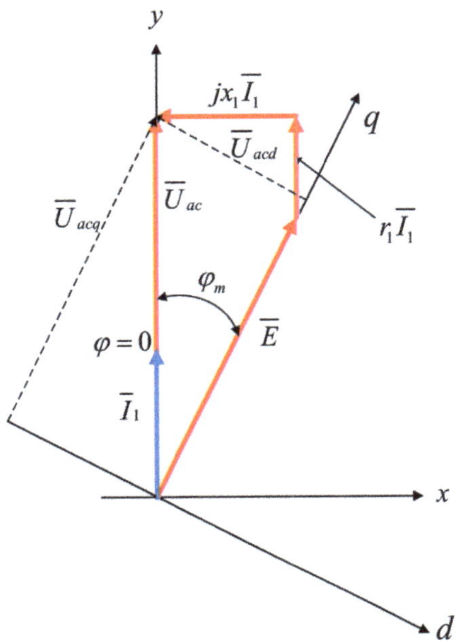

Figure 4. Vector diagram of inverter–PMSM system during energy optimization.

By performing the energy optimization (Figure 4), U_{acd}, U_{acq} control signals in d, q coordinates are set in such a way that the phase shift between the voltage vector $\overline{U}_{ac}$ and the current vector $\overline{I}_1$ at the inverter output is equal to zero ($\varphi = 0$) (Figure 4). This is equivalent to resetting the reactive power $Q_{ac} = 0$.

In the optimal mode of operation, the modulation phase can be determined on the basis of the geometric relations of the vector diagram:

$$\varphi_{m.opt} = arcsin\frac{x_1 I_1}{E} = arc\sin\frac{L_1 I_1}{\psi_0} \tag{4}$$

In order to obtain energy optimization in the inverter–PMSM system, a pair of control inputs ought to be selected in accordance with the following equations:

$$\begin{aligned} U^*_{acd} &= -U_{ac}\sin\varphi_{m.opt}, \\ U^*_{acq} &= U_{ac}\cos\varphi_{m.opt}. \end{aligned} \tag{5}$$

Currents in longitudinal (d) and transverse (q) axes can be determined on the basis of the following relations:

$$\begin{aligned} I_d &= -I_1\sin\varphi_{m.opt}, \\ I_q &= I_1\cos\varphi_{m.opt}. \end{aligned} \tag{6}$$

from which it is easy to determine the current I_1 defining the angle $\varphi_{m.opt}$ on the basis of Equation (4).

Now, having $\varphi_{m.opt}$ and generating voltages U_{ac} as an output from the PID controller (driven by the error $\Delta\omega_m = \omega_m - \omega^*_m$), Equation (5) determines the desired pair of controls $U_{acd}{}^*$ and $U_{acq}{}^*$ ensuring energy optimization in a steady state.

Note that in the dynamic (transition) states, the adopted PID-type control is far from optimal, and it is even difficult to achieve any desired specifications here.

Formally, the inverter output voltage is determined from the geometric relations of the vector diagram:

$$U_{ac} = p\omega^*_m\psi_0\cos\varphi_{m.opt} + r_1 I^*_1 \tag{7}$$

Due to the fact that energy characteristics of the PMSM propulsion can be determined from the following equations:

$$\begin{aligned} P_{ac} &= 1.5(U_{acd}I_d + U_{acq}I_q), \\ Q_{ac} &= 1.5(U_{acq}I_d - U_{acd}I_q). \end{aligned} \tag{8}$$

it is easy to check (by simple substitution) that the initial assumption (phase shift between the voltage vector $\overline{U}_{ac}$ and the current vector $\overline{I}_1$ at the inverter output is equal to zero) leads to the reset of reactive power, i.e., $Q_{ac} = 1.5(U^*_{acq}I_d - U^*_{acd}I_q) = 0$. The electromagnetic moment (T_e) in the steady state is equal to

$$T_e = \frac{2}{3}p\psi_0 I_q \tag{9}$$

The calculations were based on the data of an exemplary PMSM used in the ship's propulsion (Table 1) [31].

Table 1. Data of an exemplary PMSM used in the ship's propulsion.

Parameter	Volume (Nominal Units)	Volume (Per Unit)
Rated Mechanical Power	2.0 MW	1
Rated Apparent Power	2.2419 MVA	1
Rated Line-to-Line Voltage	690 V (rms)	1
Rated Phase Voltage	398.4 V (rms)	1
Rated Stator Current	1867.76 A (rms)	1
Rated Stator Frequency	9.75 Hz	1
Rated Power Factor	0.8921	1
Rated Rotor Speed	22.5 rpm	1
Number of Pole Pairs	26	1
Rated Mechanical Torque	848.826 kNm	1
Rated Rotor Flux Linkage	5.8264 Wb (rms)	0.896
Stator Winding Resistance, R_1	0.821 mΩ	0.00387
Moment of Inertia, J	6 kgm^2	1
d-Axis Synchronous Inductance, L_d	1.5731 mH	0.4538
q-Axis Synchronous Inductance, L_q	1.5731 mH	0.4538

Energy processes (dependencies: P_{ac} and Q_{ac} to ω_m and T_e) calculated with the use of Equations (4)–(8) are presented in Figure 5.

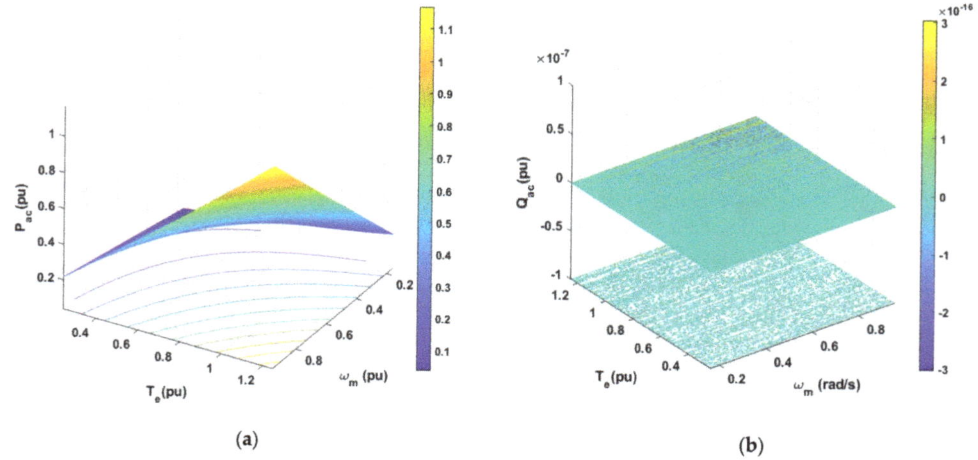

Figure 5. Energy characteristics: (**a**) active power P_{ac} and (**b**) reactive power Q_{ac} with optimization.

Analytical results of PMSM control with energy optimization shown in Figure 5 confirm the effectiveness of the proposed method. The reactive power Q_{ac} over the entire range of speed ω_m and electromagnetic torque T_e variation is zero (Figure 5b).

The electric propulsion system with PMSM, realizing energy optimization in the steady state ($Q_{ac} = 0$), is presented in Figure 6. The system includes the following:

- Voltage inverter composed of transistors (VT1–VT6) and diodes;
- Synchronous machine (PMSM);
- Rotor condition sensor (SPS);
- Coordinate transformation module (*abc/d,q*—Park–Gorev transformation);
- Coordinate transformation module (*d,q/abc*—Park–Gorev inverse transformation);
- Pulse width modulator (PWM);
- Speed controller (SC);
- Optimization block (OB).

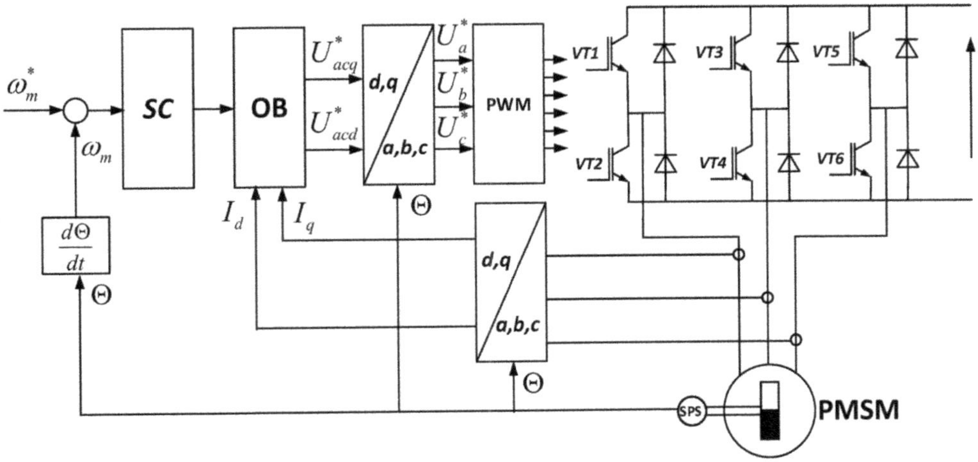

Figure 6. PMSM control system with PWM and speed regulator with energy optimization (Qac = 0).

In order to compare the control results of the PMSM system without and with energy optimization, the commonly used field-oriented control (FOC) method was implemented in MATLAB Simulink [32,33]. Figure 7 shows the energy processes in the inverter–PMSM drive system during speed and torque changes without energy optimization.

The test results of the commonly used PMSM control method (FOC), shown in Figure 7, prove that although the system responds correctly to the changes in the set speed ω^*_m and torque T_e*, the presence of non-zero reactive power Q_{ac} in the inverter–PMSM system causes additional losses (losses in the inverter, in the connecting wires, and in the machine).

The preset load torque T_L (T_e*) and the preset angular velocity are assumed to be constant only during the design process. In fact, T_L is a disturbance, acting on the system, which can be realized in different ways (it may, e.g., depend on angular velocity). All changes in T_L (or ω^*_m) can be interpreted as changes in the system equilibrium point. As the designed closed-loop system is stable, the system state just tracks the varying equilibrium point. Note that the step change of T_L (as in the paper) can be considered the strongest (worst) impact on the system.

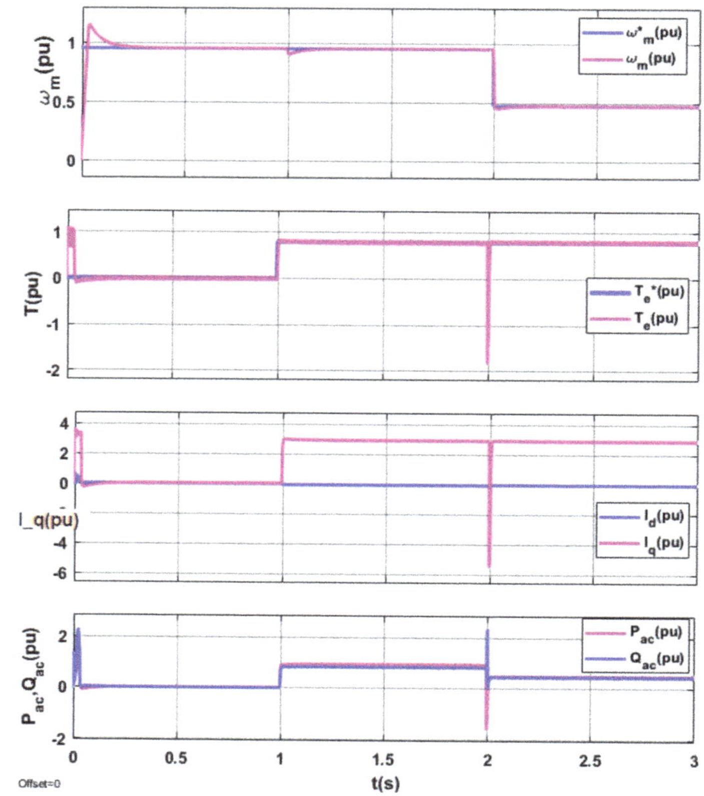

Figure 7. Energy processes (Pac, Qac) in inverter-PMSM drive system during speed and torque changes without energy optimization (FOC method).

The adopted PMSM propulsion system control strategy with energy optimization (Q_{ac} = 0) in the steady state was implemented in the MATLAB Simulink program on the basis of the model (Figure 6). PMSM parameters are shown in Table 1. During the tests, torque, rotational speed, stator currents in the d, q system, and active and reactive power during the operation of the propulsion system with energy optimization were compared. Figure 8 present the results of simulation tests. The waveforms indicate the behavior of electromagnetic torque (T_e) in relation to the set load torque T_L*, speed ω_m in relation to the set speed ω_m*, active power (P_{ac}), reactive power (Q_{ac}), and stator currents (on d and q axes) I_d and I_q in the inverter–PMSM system.

The test results of the proposed PMSM control method, shown in Figure 8, based on the analysis of steady-state electromagnetic processes, show the correct response of the system to changes in the set speed and torque (similar to the classical control method (Figure 7)) while reducing the reactive power to zero at the same time. Additional losses in the inverter–PMSM system associated with reactive power do not occur. Simulation studies confirmed the analytical research presented earlier. The proposed control method enabled energy optimization in a steady state to be obtained. In the dynamic state (t = 2 s), during the change (reduction tests) in speed, there are oscillations in the electromagnetic torque *Te*. This is connected with the fact that the synthesis of the speed controller in the classic system is conducted in accordance with the criteria of the so-called "technical optimum" [34].

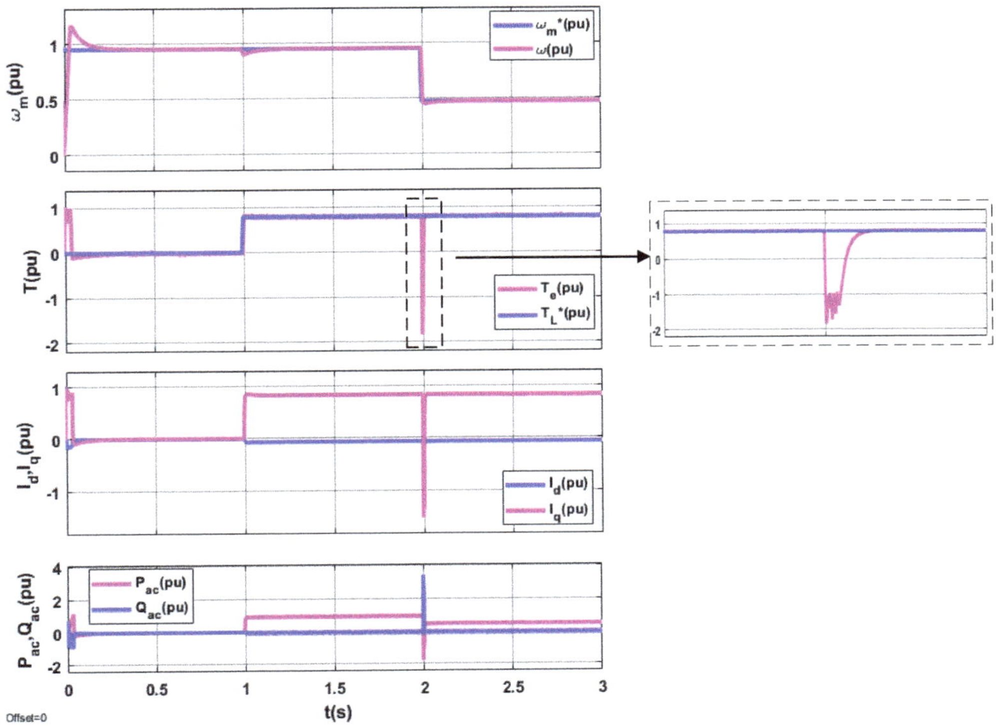

Figure 8. Energy processes (Pac, Qac) in the inverter-PMSM propulsion system during changes in speed and torque with energy optimization.

2.2. *Synthesis and Optimization of Dynamic and Static Control of a Propulsion System with PMSM*

An alternative approach to the issue, having the advantage that we can additionally obtain optimal control in dynamic (transient) states, is the use of control theory methods [35–39].

In order to optimize the operation of the propulsion system in a dynamic state, taking into account energy optimization ($Q_{ac} = 0$), an analysis with the use of input–output (I/O) linearization and optimal control techniques was carried out.

2.2.1. Design of the Main Controller via I/O Linearization Method

To facilitate the analysis and for the sake of clarity, the traditional notations in the control theory have been adopted. Let us first write Model (1) in the following state-space form:

$$\begin{aligned} \frac{dx_1}{dt} &= -\frac{r_1}{L_1}x_1 + px_3x_2 + \frac{1}{L_1}u_1 \\ \frac{dx_2}{dt} &= -px_3x_1 - \frac{r_1}{L_1}x_2 - \frac{p\psi_0}{L_1}x_3 + \frac{1}{L_1}u_2 \\ \frac{dx_3}{dt} &= \frac{1.5p\psi_0}{J}x_2 - \frac{1}{J}T_L \end{aligned} \quad (10)$$

where:

$x_1 = I_d$, $x_2 = I_q$, $x_3 = \omega_m$;

$u_1 = U_{acd}$, $u_2 = U_{acq}$.

In addition, the model parameters are the same as in Table 1; i.e., $r_1 = 0.821$ mΩ, $L_1 = 1.5731$ mH, $p = 26$, $Y_0 = 5.8264$ Wb, $J = 6$ kgm^2, $T_L = 848.826$ kNm.

The considered control problem is to stabilize the state variable x_3 on the desired x_3^* value, i.e., bring the system output

$$y = x_3 - x_3^* \tag{11}$$

to zero, by using control variables u_1 and u_2, where $x_3^* = \omega_m^*$ denotes the desired value of ω_{m}.

Analyzing the form of the above nonlinear system (10), one can observe that for controlling the output y it is enough to manipulate only the control variable u_2 while u_1 may serve us as an extra (constant) control parameter which will be then used for the system energetic optimization.

The synthesis of control u_2 can be performed by the input–output (I/O) feedback linearization, a method well known in the field of nonlinear control system engineering [27].

Based on this method, we transform Models (10) and (11) to the canonical form suitable for the direct derivation of the control law u_2.

However, to avoid the formalism of Lie derivatives, characteristic here, we repeatedly differentiate the output (11) with respect to time until the appearance of control u_2.

First we denote $y_1 = y$.

Hence, using (10), we obtain the following:

$$\dot{y}_1 = \frac{1.5p\psi_0}{J}x_2 - \frac{T_L}{J} \triangleq y_2 \tag{12}$$

Differentiating (12) once more (and again using (10)), we obtain the following:

$$\dot{y}_2 = \frac{1.5p\psi_0}{JL_1}(-pL_1x_3x_1 - r_1x_2 - p\psi_0x_3 + u_2) \tag{13}$$

Thus, we obtain the canonical form of some (sub)system of (10):

$$\begin{aligned} \dot{y}_1 &= y_2 \\ \dot{y}_2 &= f(\boldsymbol{x}) + g(\boldsymbol{x})u_2 \end{aligned} \tag{14}$$

where:

$$\begin{aligned} y_1 &= y = x_3 - x_3^*, \\ y_2 &= \dot{y}_1 = \tfrac{1.5p\psi_0}{J}x_2 - \tfrac{T_L}{J}. \end{aligned}$$

and

$$\begin{aligned} f(\boldsymbol{x}) &= \tfrac{1.5p\psi_0}{JL_1}(-pL_1x_3x_1 - r_1x_2 - p\psi_0x_3), \\ g(\boldsymbol{x}) &= \tfrac{1.5p\psi_0}{JL_1}. \end{aligned} \tag{15}$$

As the control we are looking for has the following general form:

$$u_2 = \frac{-f(\boldsymbol{x}) + k_1y_1 + k_2y_2}{g(\boldsymbol{x})} \tag{16}$$

we obtain the following specific formula:

$$u_2 = (r_1x_2 + p\psi_0x_3 + pL_1x_3x_1) + \frac{JL_1}{1.5p\psi_0}(k_1y_1 + k_2y_2) \tag{17}$$

which, to be implemented, requires measuring access to the whole system state components x_1, x_2, and x_3.

The resulting feedback system for the transformed system (10) is therefore of the following form:

$$\begin{aligned} \dot{y}_1 &= y_2 \\ \dot{y}_2 &= k_1y_1 + k_2y_2 \end{aligned}$$

Writing the last system in the matrix form

$$\begin{bmatrix} \dot{y}_1 \\ \dot{y}_2 \end{bmatrix} = \begin{bmatrix} 0 & 1 \\ 0 & 0 \end{bmatrix} \begin{bmatrix} y_1 \\ y_2 \end{bmatrix} + \begin{bmatrix} 0 \\ 1 \end{bmatrix} u \tag{18}$$

where:

$$u = \begin{bmatrix} k_1 & k_2 \end{bmatrix} \begin{bmatrix} y_1 \\ y_2 \end{bmatrix} = k_1 y_1 + k_2 y_2$$

and selecting the standard quadratic optimality criterion

$$J(u) = \int_0^{\infty} (\boldsymbol{y}^T \boldsymbol{C} \boldsymbol{y} + u \boldsymbol{D} u) dt \quad \rightarrow \min$$

where $\boldsymbol{C}$ and $\boldsymbol{D}$ are criterial matrices of proper dimensions, we can see that this controllable system can be appropriately (optimally) stabilized by the right selection of the gains k_1 and k_2 via, e.g., the LQR technique. In other words, the proper choice of these gains can ensure the required transient process of the system after a step change of certain quantities, e.g., the desired value of ω_m or the load torque T_L.

2.2.2. Analysis of the System Zero Dynamics

The partial transformation of the coordinates from x to y is defined by Equation (15). After this transformation, we have obtained a new system (14) of the second order, representing the external dynamics. As the original systems (10) and (11) are of the third order, the missing part of the dynamics should be completed with the internal (zero) dynamics (of the first order). The proof of the stability of the internal dynamics plays a key role in the successful application of the I/O linearization method, as it guarantees the stability of the whole (original) system (10).

To analyze the system zero dynamics, it is enough to examine the original systems (10) and (11) assuming that their outputs are identically equal to zero. This leads to the system's restricted motion [27] being confined to the following set:

$$\begin{aligned} Z^* &= \left\{ x : h(x) = L_f h(x) = 0 \right\} = \left\{ x : x_3 - x_3^* = 0\,;\ \tfrac{1.5p\psi_0}{J} x_2 - \tfrac{T_L}{J} = 0 \right\} \\ &= \left\{ x : x_3 = \omega_m\,;\ x_2 = \tfrac{T_L}{1.5p\psi_0} \right\} \end{aligned} \tag{19}$$

where:

$$y = h(x) = x_3 - x_3^*$$

The motion of the original systems (10) and (11) on Z^*, with the input [35]

$$u_2^* = \frac{-f(x)}{g(x)}_{|x \in Z^*} = -pL_1 x_3 x_1 - r_1 x_2 - p\psi_0 x_3$$

represents the system zero dynamics:

$$\frac{dx_1}{dt} = -\frac{r_1}{L_1} x_1 + p x_3 x_2 + \frac{1}{L_1} u_1 \tag{20}$$

As the term $p x_3 x_2$ is constant (see (19)) and u_1 is also taken as a constant parameter, this dynamics takes the form of a stable linear system:

$$\frac{dx_1}{dt} = -\frac{r_1}{L_1} x_1 + const \tag{21}$$

By choosing u_1 so that $-p x_3 x_2 = \frac{1}{L_1} u_1$, we can make the above system also asymptotically stable.

In this way, we have proved that the control (16) found above stabilizes the whole nonlinear system (10) regardless of the choice of the control u_1 (in the form of some constant parameter; compare (22)).

2.2.3. Energetic Optimization of the System Operation in Steady State

By choosing the control parameter u_1 appropriately, we may achieve some extra properties of the system (10) in the steady state. We can, e.g., perform an energetic optimization of the system operation by guaranteeing that the system reactive power is equal to zero.

Observe that by applying saturating control of the following form:

$$u_1 = a \cdot sat\left(\frac{-pL_1x_3x_2}{a}\right) \quad (22)$$

where:

$a > 0$—control parameter

we have a chance (by an appropriate choice of the parameter a) to influence the value of variable x_1 (in the steady state) without affecting the evolution (transient process) of the variables x_1 and x_2.

This way, we have the possibility to perform some steady-state system optimization.

At this point, it seems reasonable to perform an energetic optimization of the system operation in the steady state, i.e., to guarantee that the system reactive power (cf. (8)) equals zero:

$$Q_{ac} = 1.5(U_qI_d - U_dI_q) = 1.5(u_2x_1 - u_1x_2) = 0 \quad (23)$$

To analyze the system (11) in the steady state, we assume that the derivatives on the left-hand side are equal to zero.

Having access to the system state components x_1, x_2, and x_3, we can see (from the first equation of (10)) that $u_1 = r_1x_1 - pL_1x_2x_3$.

Since, in the steady state, u_1 = a (cf. (22)), we have the following:

$$x_1(a) = (pL_1x_2x_3 + a)/r_1 \quad (24)$$

Now, from the second equation of (11), we obtain the following:

$$u_2(a) = pL_1x_1(a)x_3 + r_1x_2 + p\psi_0x_3 \quad (25)$$

In summary, the reactive power in the steady state is as follows:

$$Q_{ac}(a) = 1.5(u_2x_1 - u_1x_2) = 1.5(u_2(a)x_1(a) - ax_2) \quad (26)$$

To obtain $Q_{ac}(a) = 0$, we have to just solve simple quadratic Equation (26) with respect to parameter a.

On the other hand, we can also influence the quality of the system transient process (e.g., during a load T_L step change) by appropriate selection of the gains k_1 and k_2 (cf. (18)). It should be noted that for the implementation of the controller (17) and (22), the value of the load T_L need not be known.

2.2.4. Tests of the Propulsion System with PMSM Using Synthesized Control

Based on the above-mentioned analysis, tests with the use of the MATLAB Simulink program for the DE propulsion control system of the ship (with synthesized control) were carried out. A block diagram of the linearized I/O system with energy optimization (Q_{ac}) is presented in Figure 9.

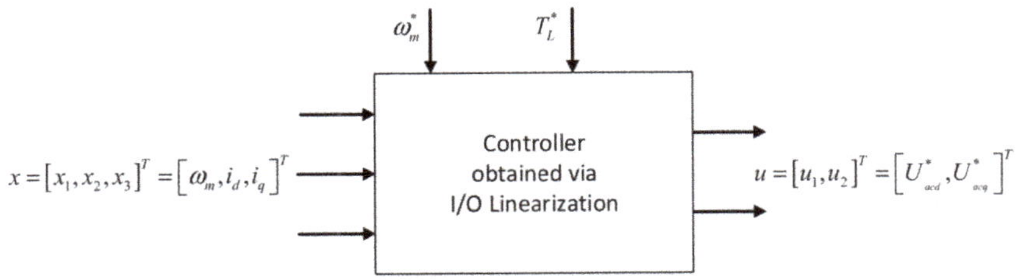

Figure 9. Block diagram of I/O linearized system with energy optimization (Q_{ac}).

Control signals U_{acd} and U_{acq}, worked out by the controller system in accordance with the strategy presented above, are setpoints for the Park–Gorev inverse transformation in the PMSM control system (Figure 6).

Simulation tests of the PMSM propulsion system with the use of the I/O linearization method allowing for dynamic optimization and energy optimization (Q_{ac}) in steady state were conducted on the basis of the model implemented in the MATLAB Simulink program. Figure 10 presents the behavior of the electromagnetic torque (T_e) in relation to the set load torque T_L*, speed ω_m in relation to the set speed ω_m*, active power (P_{ac}) and reactive power (Q_{ac}), and stator currents I_d and I_q (d and q axes) in the inverter–PMSM system.

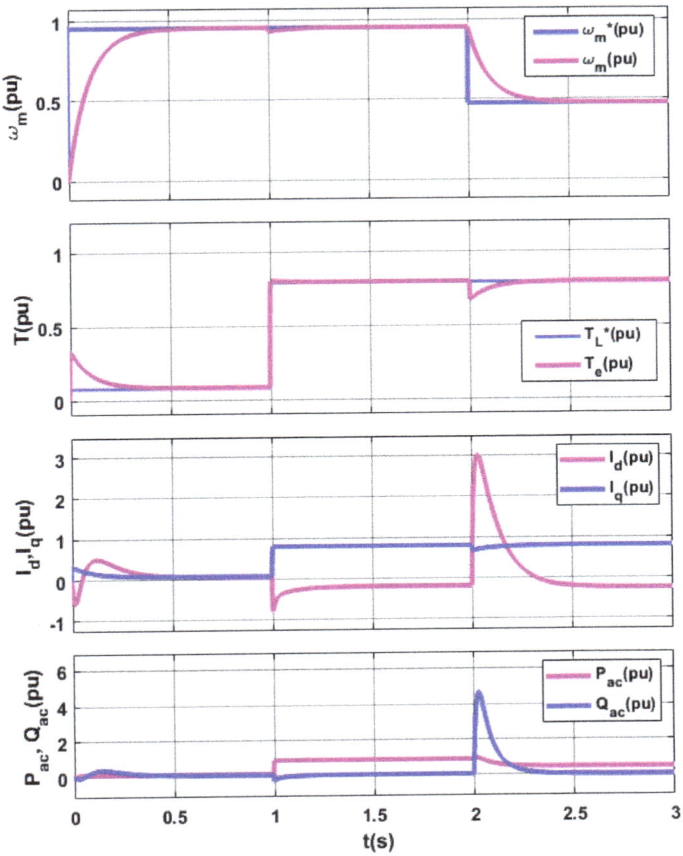

Figure 10. Energy processes (Pac, Qac) in the inverter-PMSM propulsion system during changes in speed and torque with energy optimization using the I/O linearization method.

The test results of the proposed PMSM control method shown in Figure 10 show energy optimization (Q_{ac} = 0) of the inverter–PMSM system in the steady state similarly to the first proposed method (Figure 8). At the same time, the correct response of the system in dynamic states should be noted. There is no oscillation of the moment when the speed is changed, as was the case with the first method.

3. Discussion

The article presents two methods of controlling the ship's electric propulsion. They enable obtaining energy optimization of the system by reducing the reactive power in the inverter–PMSM system (Q_{ac} = 0).

The results of analytical tests (Figure 5) confirmed in the simulation tests (Figure 8) of the modified classic method regarding the PMSM control, based on the analysis in the steady state, enabled optimizing the mechatronic propulsion system of the ship in terms of the energy criterion. Reactive power in the inverter–PMSM system has been (practically) reduced to zero. Therefore, the currents and losses in transmission lines (in the inverter and in PMSM) were limited.

The simulation results of the modified classic method of PMSM propulsion control (Figure 8) indicate the appropriate reaction of the system to the change of the set parameters ($T_L{}^*$, $\omega_m{}^*$) in the steady state. This results from the assumptions of the method. Oscillations in the electromagnetic torque T_e (t = 2 s), when changing the set speed, are connected with the lack of proper control of the system operation in the dynamic (transient) mode.

To cope with this problem, this article also proposes an alternative approach based on the methods of control theory (Section 2.2). The I/O linearization technique associated with the LQR method, from the area of optimal control, was used for a nonlinear PMSM model with energy optimization (Q_{ac} = 0).

The results of simulation tests for this control method, presented in Figure 10, demonstrate the same reaction of the system in the steady state as in the classic method (Figure 8). In the steady state, the reactive power Q_{ac} in the inverter–PMSM system is equal to zero (energy optimization). Note however that in the dynamic mode (at the change of a set speed, t = 2 s), there are no oscillations in the electromagnetic torque (T_e), as was observed in the classic control method (which copes only in the steady state).

The paper does not address the problem of knowledge of the model parameters and their errors. However, the influence of disturbances in the form of variable (step change) T_L (or $\omega^*{}_m$) has been considered. It should be noted that the approach presented in Section 2.2 allows for straightforward generalization to the adaptive version (cf. [40]) which is in preparation.

4. Conclusions

Methods of reactive power compensation based on instantaneous power analysis [41,42] in electrical systems and reactive power reduction in drive systems with PMSMs based on steady-state analysis (e.g., [43–46]) are known. Most of these methods are characterized by a complex control algorithm or a complicated system topology.

The first method proposed in the article (Section 2.1), whose simple control algorithm is based on steady-state analysis, is an effective energy optimization without significantly burdening the processor unit in the control system. The disadvantage of this method, as demonstrated by tests, is torque oscillations when dynamic changes in torque or speed occur.

The ship propulsion PMSM control strategy proposed in Section 2.2 (the second method), based on the I/O linearization technique associated with the LQR method, is effective for steady-state energy optimization. In addition, it eliminates the disadvantage of the first method by optimizing the system in transients, as studies have shown. The authors recommend this control method in ship propulsion systems where rapid changes in propulsion load occur (especially in bad weather conditions). The authors have not found a control method in the literature that integrates these two functions (steady-state energy optimization and transient optimization).

The optimal control system synthesis of the PMSM electric drive can also be used in other PMSM drives, especially in high-power drives, where it is important to reduce reactive power in the inverter–PMSM system. For drives used on land, where there are no dynamic changes in load and speed (e.g., fans in industry), the first method is recommended because the control algorithm is simple (no load on the processor unit).

Author Contributions: Conceptualization, Z.Z., D.T., S.G.-G. and M.J.; methodology, Z.Z., D.T. and S.G.-G.; validation, Z.Z., D.T. and S.G.-G.; formal analysis, Z.Z., D.T. and S.G.-G.; investigation, Z.Z., D.T. and S.G.-G.; writing—original draft preparation, Z.Z., D.T. and S.G.-G.; writing—review and editing, Z.Z., D.T. and S.G.-G. All authors have read and agreed to the published version of the manuscript.

Funding: This research received no external funding.

Institutional Review Board Statement: Not applicable.

Informed Consent Statement: Not applicable.

Data Availability Statement: Not applicable.

Conflicts of Interest: The authors declare no conflict of interest.

References

1. UNCTAD. *Review of Maritime Transport 2018*; United Nations Publication: New York, NY, USA, 2018; pp. 23–26. Available online: https://unctad.org/system/files/official-document/rmt2018_en.pdf (accessed on 5 May 2022).
2. IMO RESOLUTION MEPC.203(62) Adopted on 15 July 2011. Available online: https://wwwcdn.imo.org/localresources/en/KnowledgeCentre/IndexofIMOResolutions/MEPCDocuments/MEPC.203(62).pdf (accessed on 5 May 2022).
3. Polakis, M.; Zachariadis, P.; Kat, J.O. The Energy Efficiency Design Index (EEDI). In *Sustainable Shipping*; Springer: New York, NY, USA, 2019; pp. 93–135. [CrossRef]
4. International Maritime Organization. IMO Train the Trainer (TTT) Course on Energy Efficiency Ship Operation. In *Module 2—Ship Energy Efficiency Regulations and Related Guidelines*; International Maritime Organization: London, UK, 2016. Available online: https://wwwcdn.imo.org/localresources/en/OurWork/Environment/Documents/Air%20pollution/TTT%20trainers%20manual%20final2.pdf (accessed on 5 May 2022).
5. Grljušić, M.; Medica, V.; Radica, G. Calculation of Efficiencies of a Ship Power Plant Operating with Waste Heat Recovery through Combined Heat and Power Production. *Energies* **2015**, *8*, 4273–4299. [CrossRef]
6. Mrzljak, V.; Zarkovic, B.; Prpic-Orsic, J. Marine slow speed two-stroke diesel engine—Numerical analysis of efficiencies and im-portant operating parameters. *Mach. Technol. Mater.* **2017**, *11*, 481–484.
7. Zaccone, R.; Campora, U.; Martelli, M. Optimization of a diesel-electric ship propulsion and power generation system using a genetic algorithm. *J. Mar. Sci. Eng.* **2021**, *9*, 587. [CrossRef]
8. Hansen, J.F.; Wendt, F. History and State of the Art in Commercial Electric Ship Propulsion, Integrated Power Systems, and Future Trends. *Proc. IEEE* **2015**, *103*, 2229–2242. [CrossRef]
9. Vossen, C. Diesel electric propulsion on ∑IGMA class corvettes. In Proceedings of the 2011 IEEE Electric Ship Technologies Symposium, Alexandria, VA, USA, 10–13 April 2011; pp. 288–291. [CrossRef]
10. Acanfora, M.; Balsamo, F.; Fantauzzi, M.; Lauria, D.; Proto, D. Load levelling through Storage System for Hybrid Diesel Electric Ship Propulsion in Irregular Wave Conditions. In Proceedings of the 2022 International Symposium on Power Electronics, Electrical Drives, Automation and Motion (SPEEDAM), Sorrento, Italy, 22–24 June 2022; pp. 640–644. [CrossRef]
11. Fan, Z.; Liang, C.; Dong, Y.; Chang, G.; Pan, W.; Shao, S. Control of the Diesel-Electric Hybrid Propulsion in a Practical Cruise Vessel. In Proceedings of the 2021 IEEE International Conference on Power, Intelligent Computing and Systems (ICPICS), Shenyang, China, 29–31 July 2021; pp. 311–316. [CrossRef]
12. Veksler, A.; Johansen, T.J.; Skjetne, R.; Mathiesen, E. Thrust Allocation With Dynamic Power Consumption Modulation for Diesel-Electric Ships. *IEEE Trans. Control Syst. Technol.* **2016**, *24*, 578–593. [CrossRef]
13. Zhu, H.H.; Dong, Z. Modeling and Simulation of Hybrid Electric Ships with AC Power Bus—A Case Study. In Proceedings of the 2018 14th IEEE/ASME International Conference on Mechatronic and Embedded Systems and Applications (MESA), Oulu, Finland, 2–4 July 2018; pp. 1–8. [CrossRef]
14. WARTSILA. Improving Efficiency. Available online: https://www.wartsila.com/sustainability/innovating-for-sustainability/improving-efficiency (accessed on 5 May 2022).
15. Kozak, M. Alternating Current Electric Generator Machine Inverters in a Parallel Power Sharing Connection. *IEEE Access* **2019**, *7*, 32154–32165. [CrossRef]
16. Kozak, M. New concept of ship's power plant system with varying rotational speed gensets. In Proceedings of the 58th international conference of machine design departments (ICMD 2017), Prague, Czech Republic, 6–8 September 2017; pp. 170–175.

17. Choi, I.C.; Jeung, Y.-C.; Lee, D.-C. Variable speed control of diesel engine-generator using sliding mode control. In Proceedings of the 2017 IEEE Transportation Electrification Conference and Expo, Asia-Pacific (ITEC Asia-Pacific), Harbin, China, 7–10 August 2017; pp. 1–5. [CrossRef]
18. Rezkallah, M.; Dubuisson, F.; Singh, S.; Singh, B.; Chandra, A.; Ibrahim, H.; Ghandour, M. Coordinated Control Strategy for Hybrid off-Grid System Based on Variable Speed Diesel Generator. *IEEE Trans. Ind. Appl.* **2022**, *58*, 4411–4423. [CrossRef]
19. Hu, Y.; Cirstea, M.; McCormick, M.; Haydock, L. Modelling and simulation of a variable speed stand-alone generator system. In Proceedings of the 2000 Eighth International Conference on Power Electronics and Variable Speed Drives (IEE Conf. Publ. No. 475), London, UK, 18–19 September 2000; pp. 372–377. [CrossRef]
20. Zhou, Z.; Camara, M.B.; Dakyo, B. Coordinated Power Control of Variable-Speed Diesel Generators and Lithium-Battery on a Hybrid Electric Boat. *IEEE Trans. Veh. Technol.* **2017**, *66*, 5775–5784. [CrossRef]
21. German-Galkin, S.; Tarnapowicz, D. Energy Optimization of Ship's Shaft Generator with Permanent Magnet Synchronous Generator. *Nase More* **2020**, *67*, 138–145. [CrossRef]
22. Jamieson, P. *Innovation in Wind Turbine Design*, 2nd ed.; John Wiley & Sons, Ltd.: Hoboken, NJ, USA, 2018; ISBN 978-1-119-13790-0.
23. Ludwig, F.; Möckel, A. Operation method for AC grid powered PMSM with open-end winding in dual-inverter topology for power factor maximization. In Proceedings of the 8th IET International Conference on Power Electronics, Machines and Drives (PEMD 2016), Glasgow, UK, 19–21 April 2016; pp. 1–5. [CrossRef]
24. Jianbo, C.; Yuwen, H.; Wenxin, H.; Yong, L.; Jianfei, Y.; Mingjin, W. Direct active and reactive power control of PMSM. In Proceedings of the 2009 IEEE 6th International Power Electronics and Motion Control Conference, Wuhan, China, 17–20 May 2009; pp. 1808–1812. [CrossRef]
25. Lei, J.; Zhou, B.; Bian, J.; Qin Jiangsu, X. Unit power factor control of PMSM fed by indirect matrix converter. In Proceedings of the 2014 17th International Conference on Electrical Machines and Systems (ICEMS), Hangzhou, China, 22–25 October 2014; pp. 926–930. [CrossRef]
26. Cimini, G.; Ippoliti, G.; Orlando, G.; Pirro, M. PMSM control with power factor correction: Rapid prototyping scenario. In Proceedings of the 4th International Conference on Power Engineering, Energy and Electrical Drives, Istanbul, Turkey, 13–17 May 2013; pp. 688–693. [CrossRef]
27. Khalil, H.K. *Nonlinear Systems*, 3rd ed.; Prentice Hall: Upper Saddle River, NJ, USA, 2002; pp. 509–521; ISBN 978-0130673893.
28. Lewis, F.L.; Draguna Vrabie, D.; Syrmos, V.L. *Optimal Control*, 3rd ed.; John Wiley & Sons, Inc.: Hoboken, NJ, USA, 2012; ISBN 978-0470633496.
29. German-Galkin, S.; Tarnapowicz, D. Energy Optimization of the 'Shore to Ship' System-A Universal Power System for Ships at Berth in a Port. *Sensors* **2020**, *20*, 3815. [CrossRef] [PubMed]
30. German-Galkin, S.; Tarnapowicz, D. Optimization of the electric car's drive system with PMSM. In Proceedings of the 12th International Science-Technical Conference AUTOMOTIVE SAFETY, Kielce, Poland, 21–23 October 2020. [CrossRef]
31. Wu, B.; Lang, Y.; Zargari, N.; Kouro, S. Appendix B: Generator Parameters. In *Power Conversion and Control of Wind Energy Systems*; IEEE: Manhattan, NY, USA, 2011; pp. 319–326. [CrossRef]
32. Zawirski, K. *Permanent Magnet Synchronous Motor Control*; Poznan University of Technology Publishing House: Poznań, Poland, 2005; ISBN 83-7143-337-9.
33. Kaczmarek, T.; Zawirski, K. *Drive Systems with Synchronous Motor*; Poznan University of Technology Publishing House: Poznań, Poland, 2000; ISBN 978-83-7348-401-6.
34. Kazmierkowski, M.P.; Tunia, H. Automatic Control of Converter-Fed Drives. In *Studies in Electrical and Electronic Engineering*, 1st ed.; PWN-Elsevier Science Publishers: Warsaw, Polland; Amsterdam, The Netherlands; New York, NY, USA; Tokyo, Japan, 1994; ISBN 13-978-0444986603. (In English)
35. Aghili, F. Optimal Feedback Linearization Control of Interior PM Synchronous Motors Subject to Time-Varying Operation Conditions Minimizing Power Loss. *IEEE Trans. Ind. Electron.* **2018**, *65*, 5414–5421. [CrossRef]
36. Fraga, R.; Sheng, L. Non linear and intelligent controllers for the ship rudder control. In Proceedings of the 2011 Saudi International Electronics, Communications and Photonics Conference (SIECPC), Riyadh, Saudi Arabia, 24–26 April 2011; pp. 1–5. [CrossRef]
37. Parvathy, A.K.; Devanathan, R.; Kamaraj, V. Application of quadratic linearization to control of Permanent Magnet synchronous motor. In Proceedings of the 2011 1st International Conference on Electrical Energy Systems, Chennai, India, 3–5 January 2011; pp. 158–163. [CrossRef]
38. Do, T.D.; Kwak, S.; Choi, H.H.; Jung, J.-W. Suboptimal Control Scheme Design for Interior Permanent-Magnet Synchronous Motors: An SDRE-Based Approach. *IEEE Trans. Power Electron.* **2014**, *29*, 3020–3031. [CrossRef]
39. Cheema, M.A.M.; Fletcher, J.E.; Xiao, D.; Rahman, M.F. A Linear Quadratic Regulator-Based Optimal Direct Thrust Force Control of Linear Permanent-Magnet Synchronous Motor. *IEEE Trans. Ind. Electron.* **2016**, *63*, 2722–2733. [CrossRef]
40. Zwierzewicz, Z. Robust and Adaptive Path-Following Control of an Underactuated Ship. *IEEE Access* **2020**, *8*, 120198–120207. [CrossRef]
41. Wang, F.; Mi, Z.; Yang, Q. Comparison between Periodic Function space based power theory and traditional power theory. In Proceedings of the 2008 International Conference on Electrical Machines and Systems, Wuhan, China, 17–20 October 2008; pp. 2212–2216.
42. Akagi, H.; Hirokazu-Watanabe, E.; Aredes, M. The Instantaneous Power Theory. In *Instantaneous Power Theory and Applications to Power Conditioning*; IEEE: Manhattan, NY, USA, 2017; pp. 37–109. [CrossRef]

43. Ho, T.-Y.; Chen, M.-S.; Chen, L.-Y.; Yang, L.-H. The design of a PMSM motor drive with active power factor correction. In Proceedings of the 2011 2nd International Conference on Artificial Intelligence, Management Science and Electronic Commerce (AIMSEC), Dengfeng, China, 8–10 August 2011; pp. 4447–4450. [CrossRef]
44. Tau, J.-H.; Tzou, Y.-Y. PFC control of electrolytic capacitor-less PMSM drives for home appliances. In Proceedings of the 2017 IEEE 26th International Symposium on Industrial Electronics (ISIE), Edinburgh, UK, 19–21 June 2017; pp. 335–341. [CrossRef]
45. Gokulapriya, R.; Pradeep, J. Shunt based active power factor correction circuit for direct torque controlled PMSM drive. In Proceedings of the 2017 Third International Conference on Science Technology Engineering & Management (ICONSTEM), Chennai, India, 23–24 March 2017; pp. 517–521. [CrossRef]
46. Vimal, M.; Sojan, V. Vector controlled PMSM drive with power factor correction using zeta converter. In Proceedings of the 2017 International Conference on Energy, Communication, Data Analytics and Soft Computing (ICECDS), Chennai, India, 1–2 August 2017; pp. 289–295. [CrossRef]

energies

MDPI

Article

Interturn Short Fault Detection and Location of Permanent Magnet Wind Generator Based on Negative Sequence Current Residuals

Tonghua Wu [1,2,3], Shouguo Cai [3,*], Wei Dai [1,2], Ying Zhu [3,*], Xiaobao Liu [1] and Xindong Li [1]

1 NARI Group Corporation, Nanjing 211106, China
2 State Key Laboratory of Smart Grid Protection and Control, Nanjing 211106, China
3 College of Energy and Electrical Engineering, Hohai University, Nanjing 210098, China
* Correspondence: sgcai@hhu.edu.cn (S.C.); yingzhu@hhu.edu.cn (Y.Z.)

Abstract: This article proposes a model-based method for the detection and phase location of interturn short fault (ISF) in the permanent magnet synchronous generator (PMSG). The simplified mathematical model of PMSG with ISF on *dq*-axis is established to analyze the fault signature. The current residuals are accurately estimated through Luenberger observer based on the expanded mathematical model of PMSG. In current residuals, the second harmonics are extracted using negative sequence park transform and angular integral filtering to construct the fault detection index. In addition, the unbalance characteristics of three-phase current after ISF can reflect the location of the fault phase, based on which the location indexes are defined. Simulation results for various operating and fault severity conditions primarily validate the effectiveness and robustness of diagnosis method in this paper.

Keywords: permanent magnet synchronous generator; interturn short fault; current residuals; negative sequence; fault detection; fault phase location

Citation: Wu, T.; Cai, S.; Dai, W.; Zhu, Y.; Liu, X.; Li, X. Interturn Short Fault Detection and Location of Permanent Magnet Wind Generator Based on Negative Sequence Current Residuals. *Energies* **2022**, *15*, 9441. https://doi.org/10.3390/en15249441

Academic Editor: Gianluca Brando

Received: 21 November 2022
Accepted: 12 December 2022
Published: 13 December 2022

1. Introduction

Permanent magnet synchronous machines (PMSMs) have the advantages of high power density, high efficiency and stability, so it has a wide range of applications in aerospace, electric automobile and new energy power generation [1,2]. In terms of wind power generation, permanent magnet synchronous generators (PMSGs) are currently used extensively compared to doubly fed induction generators (DFIGs), especially in direct-driven power generation systems (PGs) for low weight and volume and the capability of multipole design of PMSGs [3–5].

Interturn short faults (ISFs), demagnetization faults, rotor eccentricity and other faults are common faults during the operation of PMSGs, among which ISFs occur most frequently [6]. Due to the damp environment, mechanical vibration, instantaneous overvoltage and other reasons, the winding insulation of PMSGs may be damaged, resulting in the occurrence of ISFs [7]. ISFs will produce large fault current both in the short circuit and fault phase winding, which may cause serious secondary faults such as open circuit, interphase short circuit, and grounding fault [8]. In addition, short-circuited coils will generate magnetomotive force opposite to other coils. The direction of the magnetic field generated by the short-circuited coil is opposite to the direction of the air gap composite magnetic field, which will create the irreversible demagnetization of permanent magnets [9]. Thus, prompt and reliable detection and location algorithms will help in early fault diagnosis to conduct repair and maintenance as soon as possible.

At present, most of the research objects of fault detection technology, at home and abroad, are induction motors (IMs). Due to the relatively short time of appearance, the fault detection technology of PMSMs or PMSGs is still in the development stage. The existing fault detection methods for PMSMs can be roughly divided into the following three types: the method based on analytical model; the method based on knowledge; and the method based on signal processing [10,11]. The first method constructs an accurate mathematical model of the motor through the logical relationships between the parameters of the system, so the model can predict the output under healthy conditions. This method judges the occurrence of the fault by comparing the residuals between the predicted output and the actual output of the motor [12]. The main advantage of this approach is that no additional hardware is required to implement fault detection and isolation algorithms. In reference [13], an efficient, simplified physical faulty model considering the disposition of coils under the same or different pole-pair is proposed, which can simulate different types of stator faults. In reference [14], model predictive control (MPC) is used in the PMSM control system and detection is acquired by analyzing the dc component and the second harmonic in the cost function.

The knowledge-based method can be divided into model-based reasoning, neural network based and fuzzy logic based fault diagnosis methods, based on different sources of knowledge learning [15,16]. Common diagnosis methods include artificial neural network, fuzzy logic diagnosis, expert system and information fusion diagnosis, etc. [17]. The method avoids the high dependence on the machine model, and can introduce many aspects of motor information, which opens up a new path for research of fault detection. In reference [18], a diagnostic system based on convolutional neural network (CNN) is proposed to detect the PM damages using raw phase current signals. However, the complexity of the algorithm reduces the possibility of real-time diagnosis. To detect and locate the open-circuit of PWM-VSI in the PMSM drive, the fuzzy logic method is used to process fault diagnosis variables from the average current Park's vector in reference [19], but the problem of poor self-learning still exists. In reference [20], the vibrations and currents of the fault machine are calculated using a multidimensional diagnosis methodology, and support vector machine (SVM) is used to classify single and combined fault scenarios under different speed and torque conditions.

The signal processing method collects the voltage, current, output torque, back electromotive force (EMF) and other signals of the fault machine, then uses various signal processing methods to extract and analyze the time-frequency domain characteristics of the fault signal for fault diagnosis [21,22]. Common signal processing methods include fast Fourier transform (FFT), Hilbert-Huang transform (HHT), wavelet transform (WT) and empirical mode decomposition (EMD). The method does not need accurate motor analytical model and related parameters, and has strong adaptability, which is a mature and widely used method in motor fault detection. In reference [23], HF rotating square-waveform voltages are injected to detect and classify the turn fault and high-resistance connection (HRC) fault of PMSM, by extracting HF currents consisting of fault characteristics. Reference [24] proposes a diagnosis algorithm based on discrete wavelet transform (DWT) and SVM, and designs an adaptive filter to address the problem of harmonic removal. Reference [25] uses EMD energy entropy and normalized average current to diagnose open-circuit faults of one or more phases in the PMSM drive circuit, which can also be used for interturn fault detection.

In addition, the fast location of the fault phase ensures the efficiency and accuracy of the overall fault diagnosis algorithm of PMSMs. Common fault location algorithms, on the one hand, use the amplitude and phase difference of fault phase current or voltage as the diagnosis basis [26], or acquire the waveform distortion of the magnetomotive force under fault conditions, which require additional equipment investment [27]. Thus, there are still a lack of clear indexes for fault location under variable conditions and the accuracy of the location algorithm needs to be further improved.

The accuracy of residual-based fault diagnosis methods depends on how accurately the residual is measured. The residuals (current, voltage, EMF) can be obtained by building an accurate model of the normal machine, which can be established from lumped parameters and finite element (FE) analysis [28], or using a virtual observer such as the Luenberger observer, sliding mode observer (SMO) and extended state observer. Since data in a healthy state are generally obtained from steady-state conditions, the effects of transient speed and load conditions on the detection performance cannot be neglected. However, the measurement method based on a virtual observer can be applied to variable operating conditions without additional equipment investment.

The paper proposes a new detection and location strategy for ISF in PMSGs based on extracting and analyzing the negative sequence residuals. The current residuals are accurately measured by the Luenberger observer. The defined fault diagnosis and phase location indexes can be extracted from the residuals through negative sequence park transform and angular integral filtering. The indexes have good robustness to speed and load fluctuation, which can adapt to the speed variations of wind turbine under the maximum power point tracking (MPPT) or limited power state.

The paper consists of five essential parts. In Section 2, a simplified ISF model of PMSG on *dq*-axis is established to analyze the fault components on *dq*-axis. In Section 3, the Luenberger observer based on expanded mathematical model on *dq*-axis is designed to accurately measure current residuals under fault conditions. Section 4 defines the fault detection and fault phase location indexes, which consider speed and load fluctuation during operation of PMSG. Finally, simulation results in Matlab/Simulink are given in Section 5.

2. Mathematical Model of PMSG with ISF

2.1. Wind Turbine Aerodynamic Model

According to Baez theory and aerodynamics, the mechanical power P_w captured by wind turbines from wind energy are expressed as [29]:

$$P_w = \frac{1}{2}\pi\rho R^2 v^3 C_p(\lambda, \beta) \tag{1}$$

where ρ is the air density, R is the wind turbine radius, $\lambda = \omega_m R/v$ is the tip speed ratio, v is the wind speed, ω_m is the mechanical speed of the wind turbine; C_p is the power coefficient, β is the blade pitch.

Figure 1 shows the wind turbine power-rotor speed characteristic curves at different wind speeds, the maximum power points (MPPs) under different wind speeds forms the optimal power curve. PMSGs usually operates under maximum power point tracking (MPPT) control, where there is a optimal value λ_{opt} at which the PMSGs can extracts the maximum power P_{w_opt} from wind given by:

$$P_{w\_\max} = \frac{\pi\rho R^5 C_{pmax}\omega_m^3}{2\lambda_{opt}^3} = k\omega_m^3 \tag{2}$$

where C_{pmax} is the maximum power coefficient, k is the optimal power constant, which is only related to the characteristic parameters of the wind turbine. Accordingly, the optimal torque at MPPs is expressed as:

$$T_{w\_\max} = \frac{P_{w\_\max}}{\omega_m} = k\omega_m^2 \tag{3}$$

2.2. Modeling of the PMSG with ISF

Considering there is only one branch in each phase winding, the electrical model of PMSG with ISF in phase a is shown in Figure 2. The faulty winding a is divided into the healthy and faulted parts. The faulted part is represented by a winding with resistance R_{af}, inductance L_{af}, and back EMF e_{af}, and the current flow through it is defined as faulted current i_{af}. The contract resistance in shorted path is defined as R_f, with the short-circuit current i_f flowing through it. The ratio of number of turns in faulted winding to the total number of turns per phase is defined as μ, which ranges from 0 to 1. The ratio μ and R_f collectively represent the fault severity of the ISF. Under ideal conditions, the resistance and back EMF can be considered proportional to number of turns, and the inductance is proportional to the square of turns. Thus, the parameters both in the healthy and faulted parts can be calculated accordingly.

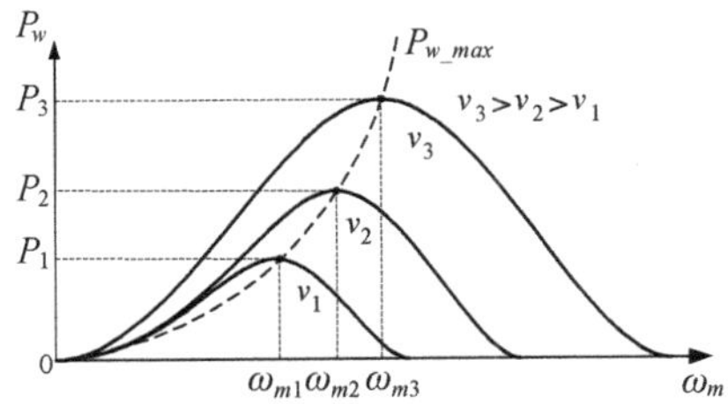

Figure 1. Wind turbine power-rotor speed characteristic curves at different wind speeds.

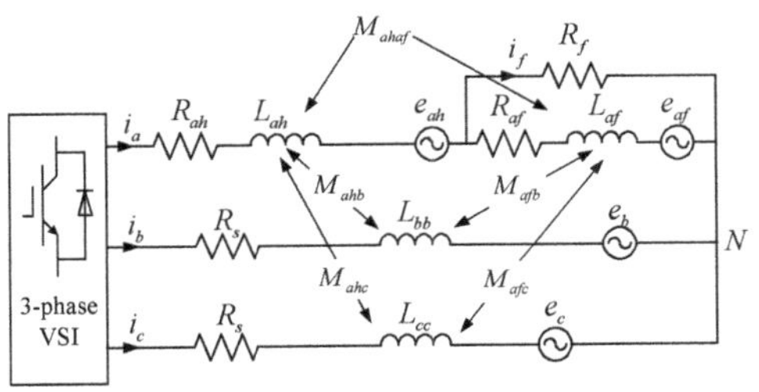

Figure 2. Electrical model of PMSG with ISF in phase *a*.

From reference [30], the mathematical model of PMSG with ISF can be expressed in (4), where R_s is stator resistance, θ_e is electrical angle. L and M are self and mutual inductance of stator winding respectively, λ_{PM} is the PM flux linkages.

$$\begin{bmatrix} u_{ah} \\ u_b \\ u_c \\ u_{af} \end{bmatrix} = \begin{bmatrix} (1-\mu)R_s & & & \\ & R_s & & \\ & & R_s & \\ & & & \mu R_s \end{bmatrix} \begin{bmatrix} i_a \\ i_b \\ i_c \\ i_{af} \end{bmatrix} + \frac{d\lambda_{PM}}{dt} \begin{bmatrix} (1-\mu)\cos\theta_e \\ \cos(\theta_e - 2\pi/3) \\ \cos(\theta_e + 2\pi/3) \\ \mu\cos\theta_e \end{bmatrix} + \begin{bmatrix} (1-\mu)^2 L_{aa} & (1-\mu)M_{ab} & (1-\mu)M_{ac} & \mu(1-\mu)L_{aa} \\ (1-\mu)M_{ba} & L_{bb} & M_{bc} & \mu M_{ba} \\ (1-\mu)M_{ca} & M_{cb} & L_{cc} & \mu M_{ca} \\ \mu(1-\mu)L_{aa} & \mu M_{ab} & \mu M_{ac} & \mu^2 L_{aa} \end{bmatrix} \frac{d}{dt} \begin{bmatrix} i_a \\ i_b \\ i_c \\ i_{af} \end{bmatrix} \tag{4}$$

Considering the voltage in the faulty phase can be written as:

$$u_a = u_{ah} + u_{af} \tag{5}$$

The current and voltage of faulted portion have relations as follows:

$$\begin{cases} i_{af} = i_a - i_f \\ u_{af} = R_f i_f \end{cases} \tag{6}$$

Using (5) and (6) in (4), the three-phase voltages equations under turn fault conditions can be rewritten as:

$$\begin{bmatrix} u_a \\ u_b \\ u_c \end{bmatrix} = \begin{bmatrix} R_s & & \\ & R_s & \\ & & R_s \end{bmatrix} \begin{bmatrix} i_a \\ i_b \\ i_c \end{bmatrix} + \begin{bmatrix} L_{aa} & M_{ab} & M_{ac} \\ M_{ba} & L_{bb} & M_{bc} \\ M_{ca} & M_{cb} & L_{cc} \end{bmatrix} \frac{d}{dt} \begin{bmatrix} i_a \\ i_b \\ i_c \end{bmatrix} + \frac{d\lambda_{pm}}{dt} \begin{bmatrix} \cos\theta_e \\ \cos(\theta_e - 2\pi/3) \\ \cos(\theta_e + 2\pi/3) \end{bmatrix} - \mu R_s \begin{bmatrix} i_f \\ 0 \\ 0 \end{bmatrix} - \mu \begin{bmatrix} L_{aa} \\ M_{ab} \\ M_{ac} \end{bmatrix} \frac{di_f}{dt} \tag{7}$$

By Park transformation, the voltage equations of PMSG with ISF in dq coordinate can be written as:

$$\begin{cases} u_{df} = R_s i_d + L_d \frac{di_d}{dt} - \omega_e L_q i_q - \frac{2}{3} R_s \mu i_f \cos\theta_e \\ \quad -\frac{2}{3}\omega_e L_q \mu i_f \sin\theta_e - \frac{2}{3} L_q \mu \cos\theta_e \frac{di_f}{dt} + \frac{2}{3}\omega_e L_d \mu i_f \sin\theta_e \\ u_{qf} = R_s i_q + L_q \frac{di_q}{dt} + \omega_e L_d i_d + \omega_e \lambda_{PM} + \frac{2}{3} R_s \mu i_f \sin\theta_e \\ \quad -\frac{2}{3}\omega_e L_d \mu i_f \cos\theta_e + \frac{2}{3} L_q \mu \sin\theta_e \frac{di_f}{dt} + \frac{2}{3}\omega_e L_d \mu i_f \cos\theta_e \end{cases} \tag{8}$$

where ω_e is the electrical angular speed, u_{df} and u_{qf} are dq-axis voltages in ISF condition, i_d and i_q are dq-axis currents in normal condition, and L_d and L_q are dq-axis inductance. The angle θ_e will be replaced with $(\theta_e - 2\pi/3)$ or $(\theta_e + 2\pi/3)$ when ISF occurs in phase b or c.

Rearrange the voltage Equation (8) as:

$$\begin{cases} u_{df} = R_s i_{df} + L_d \frac{di_{df}}{dt} - \omega_e L_q i_{qf} \\ u_{qf} = R_s i_{qf} + L_q \frac{di_{qf}}{dt} + \omega_e (L_d i_{df} + \lambda_{PM}) \end{cases} \tag{9}$$

where i_{df} and i_{qf} are dq-axis current in ISF situation, which can be written as:

$$\begin{cases} i_{df} = i_d - \frac{2}{3}\mu i_f \cos\theta_e = i_d + \Delta i_d \\ i_{qf} = i_q + \frac{2}{3}\mu i_f \sin\theta_e = i_q + \Delta i_q \end{cases} \tag{10}$$

Equation (9) has the same form as the normal voltage equations. It means if the same voltage u_{df} and u_{qf} in (8) are applied in the normal model conditions, the fault current with ISF can be extracted and expressed in (10). The Δi_d and Δi_q are defined as current residuals, whose magnitude only depends on μ and i_f. Since the residuals equal to zero under normal conditions and can indicate the severity of the ISF, it is a stable and robust signature for ISF detection.

From reference [31], the electromagnetic torque of the PMSG with ISF is expressed as:

$$T_{ef} = \frac{e_a i_a + e_b i_b + e_c i_c - e_f i_f}{\omega_e} \tag{11}$$

where e_{abc} are the back electromotive force, the mechanical torque of the wind turbine is the load torque of PMSG, so the mechanical part of PMSG can be described as:

$$J\frac{d\omega_m}{dt} = T_{ef} - T_w - B\omega_m \tag{12}$$

where J is the rotor inertia, B is the damping coefficient, which characterizes the combined viscous friction of rotor and load.

3. Estimation of Current Residuals

3.1. Design of State Observer

As expressed in (10), the current residuals can be obtained by measuring the difference between currents in ISF and healthy conditions. However, the healthy model needs to be accurate enough to meet diagnostic needs, and this measuring method need to store and invoke voltage and current data under different operating conditions. In contrast, the flexibility of the virtual observer and the need for no additional equipment investment make up for this defect. The paper use the Luenberger observer for the estimation of current residuals for its dynamic response performance and estimation accuracy.

According to (9) and (10), the current residuals can be defined as state variables, and the expanded mathematical model of PMSG with ISF can be written as:

$$\begin{cases} \dot{x} = \boldsymbol{A}x + \boldsymbol{B}u \\ y = \boldsymbol{C}x \end{cases} \tag{13}$$

where

$$A = \begin{bmatrix} -\frac{R_s}{L_d} & \frac{\omega_e L_q}{L_d} & -\frac{R_s}{L_d} & \frac{\omega_e(L_q-2L_d)}{L_d} \\ -\frac{\omega_e L_q}{L_d} & -\frac{R_s}{L_q} & \frac{\omega_e(2L_q-L_d)}{L_q} & -\frac{R_s}{L_q} \\ 0 & 0 & 0 & 2\omega_e \\ 0 & 0 & -2\omega_e & 0 \end{bmatrix} \tag{14}$$

$$B = \begin{bmatrix} 0 & \frac{1}{L_q} & 0 & 0 \\ \frac{1}{L_d} & 0 & 0 & 0 \end{bmatrix}^T \tag{15}$$

$$C = \begin{bmatrix} 1 & 0 & 0 & 0 \\ 0 & 1 & 0 & 0 \end{bmatrix} \tag{16}$$

and $x = \begin{bmatrix} i_d & i_q & \Delta i_d & \Delta i_q \end{bmatrix}^T$ are the state variables, $y = \begin{bmatrix} i_d & i_q \end{bmatrix}^T$ are the output variables, $u = \begin{bmatrix} u_d & u_q - \omega_e \lambda_{PM} \end{bmatrix}^T$ are the control variables, A is the state matrix, B is the input matrix, and C is the output matrix.

In order to measure the current residuals, the classical Luenberger observer can be designed as:

$$\begin{cases} \dot{\hat{x}} = A\hat{x} + Bu + L(y - \hat{y}) \\ \hat{y} = C\hat{x} \end{cases} \tag{17}$$

where superscript '^' indicates the estimated value. $L = \begin{bmatrix} l_1 & l_3 & l_5 & l_7 \\ l_2 & l_4 & l_6 & l_8 \end{bmatrix}^T$ is the observation matrix. Figure 3 shows the structure diagram of the Luenberger observer.

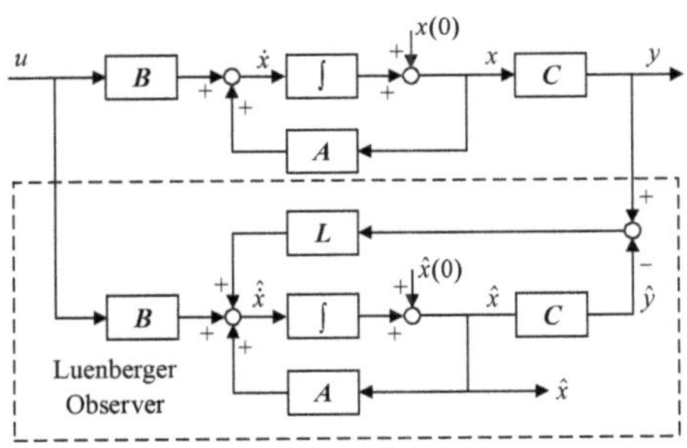

Figure 3. Structure diagram of the Luenberger observer.

The Luenberger observer reflects the error between the observation state and the actual state to the observation output through the observation matrix L. Therefore, the selection of the parameter L will affect the convergence and accuracy of the observer.

3.2. Stability Aanlysis

As mentioned above, the parameter design of observation matrix L is crucial, to simplify the parameter design process, it is set that $l_2 = l_3 = l_6 = l_7 = 0$. The state error equation of the observer can be obtained by making a difference between Equations (13) and (17) above:

$$\begin{aligned} \dot{e} = \dot{x} - \dot{\hat{x}} &= A(x - \hat{x}) - L(y - \hat{y}) \\ &= (A - LC)e \end{aligned} \tag{18}$$

where $(A - LC)$ is the system state matrix of the observer, whose eigenvalue is called system poles of the observer. When the poles of the system are all located in the left half plane of the s-domain, or the real components of the eigenvalues of state matrix $(A - LC)$ are all negative, the system response will decay and tend to be stable. The characteristic equation for $(A - LC)$ can be written as:

$$|sI - (A - LC)| = a_4 s^4 + a_3 s^3 + a_2 s^2 + a_1 s + a_0 \tag{19}$$

where

$$\begin{cases} a_4 = 1,\ a_3 = \frac{2R_s}{L_d} + l_1 + l_4 \\ a_2 = \frac{R_s^2 + \omega_e^2 L_q^2}{L_d^2} + 4\omega_e^2 + l_1 l_4 + \frac{R_s}{L_d}(l_1 + l_4 + l_5 + l_8) \\ a_1 = -\frac{(l_5+l_8)R_s^2}{L_d^2} - \frac{l_8(2L_d^3+L_q^3)\omega_e^2}{L_d^2 L_q} + 4(l_1 + l_4 + l_5 + l_8)\omega_e^2 \\ \quad + \frac{8R_s(\omega_e^2 - l_1 l_8 - l_4 l_5)}{L_d} \\ a_0 = \frac{4L_q^2\omega_e^4 + 4R_s^2\omega_e^2 + l_5 l_8 R_s^2 + 2l_8 L_q R_s \omega_e^2}{L_d^2} - \frac{2l_8 R_s \omega_e^2 + 2(l_1+l_5)l_8 L_d \omega_e^2}{L_q} \\ \quad +4(l_1 + l_5)(l_4 + l_8)\omega_e^2 - \frac{2L_q(l_4 l_5 + l_5 l_8)\omega_e^2 - 4(l_1+l_4+l_5+l_8)R_s\omega_e^2}{L_d} \end{cases} \tag{20}$$

For the observer system, the commonly used stability judgment methods include Routh criterion, Hurwitz criterion, Nyquist criterion and Root Locus method, etc. The Hurwitz criterion is more suitable for the stability analysis of the fourth order system for its simple rules and convenient use. Therefore, the Hurwitz determinant of characteristic Equation (21) can be written as:

$$D_4 = \begin{vmatrix} a_3 & a_1 & 0 & 0 \\ a_4 & a_2 & a_0 & 0 \\ 0 & a_3 & a_1 & 0 \\ 0 & a_4 & a_2 & a_0 \end{vmatrix} \tag{21}$$

According to the Hurwitz criterion, the Luenberger observer of PMSG designed above has global stability only when the principal determinant of each order of D_4 is greater than zero. In order to ensure that the observer system has a higher bandwidth than the original system without observer, the larger the parameters of the $\boldsymbol{L}$ are, the better the dynamic characteristics of the observer. However, the parameters of $\boldsymbol{L}$ are firstly limited by the saturation characteristics of the devices in actual system. Secondly, there are usually interference and measurement noise in the measured output y of the actual system. Excessive parameter setting will amplify system interference and noise. Considering the level of noise measured by the actual system, in order to avoid the pulsation caused by amplified noise, the observer poles are generally 2~10 times of the original system poles [32]. How to select the gain parameters according to the working conditions is still a key issue.

4. Detection and Location Strategy of ISF

4.1. Fault Detection

Without considering high-order harmonics, the short-circuit current i_f can be expressed as:

$$i_f \approx I_f \sin(\theta_e + \theta_F) \tag{22}$$

where I_f and θ_F are the amplitude and initial phase angle of short-circuit current. The amplitude is mainly affected by shorted turn ratio μ and fault resistance R_f, but also related to speed and load condition of PMSG. While the phase angle θ_F is associate with the location of the fault, which is synchronized with the initial phase angle of the fault phase.

By substituting (22) into (10), it has:

$$\begin{cases} \Delta i_d = -\frac{2}{3}\mu I_f \sin(\theta_e + \theta_F)\cos\theta_e \\ \quad = -\frac{1}{3}\mu I_f \sin(2\theta_e + \theta_F) - \frac{1}{3}\mu I_f \sin\theta_F \\ \quad = \Delta i_{d\_2nd} \sin(2\theta_e + \theta_F) + \Delta i_{d\_dc} \\ \Delta i_q = \frac{2}{3}\mu I_f \sin(\theta_e + \theta_F)\sin\theta_e \\ \quad = -\frac{1}{3}\mu I_f \cos(2\theta_e + \theta_F) + \frac{1}{3}\mu I_f \cos\theta_F \\ \quad = \Delta i_{q\_2nd} \cos(2\theta_e + \theta_F) + \Delta i_{q\_dc} \end{cases} \tag{23}$$

It can be concluded from (23) that the current residuals on dq-axis contain dc components and second harmonics. The amplitudes of both are related to the amplitude of short-circuit current and shorted turn ratio, both can be used as index parameters to characterize the severity of ISF. However, the dc component is also affected by the θ_F. When the θ_F approaches a certain value, the dc component residuals on axis d or q may decrease to zero, thus affecting fault judgment. In addition, when the actual parameters of the generator do not match the theoretical parameters, or when the sudden change of the speed or load torque of the motor causes the offset of the fault angle, these will introduce the dc component into the current differential Equation (9) without affecting the higher harmonic component. Therefore, this paper selects the second harmonics of current residuals on dq-axis to construct the fault index.

$$FI_{sum} = \sqrt{(\Delta i_{d\_2nd})^2 + (\Delta i_{q\_2nd})^2} \tag{24}$$

From reference [33], the amplitude of short-circuit current I_f is approximately proportional to the speed of the motor. Therefore, the above-mentioned index FI_{sum} will fluctuate when the motor speed changes or the outside wind speed changes for wind turbines, which is not conducive to the determination of the fault detection threshold. In order to eliminate the influence of the speed on the fault index, the fault index need to be reconstructed into the following form:

$$FI = \frac{\mu I_f}{\omega_e} = \frac{3\sqrt{2FI_{sum}}}{2\omega_e} \tag{25}$$

The above index is only proportional to the ratio μ, but not to other factors such as speed. The construction of FI needs to extract the second harmonics of current residuals. A common frequency domain analysis algorithm, such as FFT, HHT and WT, decompose and reconstruct signals in frequency domain or time-frequency domain. However, these algorithms need to track the frequency of the motor in real time, which will fluctuate under ISF, thus resulting in the final measurement bias. To avoid this, extraction of second harmonic residuals can be indirectly accomplished by negative sequence Park transformation as follows:

$$\begin{cases} \Delta i_d^n = \Delta i_d \cos 2\theta_e - \Delta i_q \sin 2\theta_e \\ \quad\;\; = \Delta i_{d\_dc}^n + \Delta i_{d\_2nd}^n \sin(2\theta_e + \theta_F) \\ \Delta i_q^n = \Delta i_d \sin 2\theta_e + \Delta i_q \cos 2\theta_e \\ \quad\;\; = \Delta i_{q\_dc}^n - \Delta i_{q\_2nd}^n \cos(2\theta_e + \theta_F) \end{cases} \tag{26}$$

where superscript 'n' indicates the value in negative sequence. Through the above transformation, the second harmonics of the current residuals in the positive sequence coordinates are the dc components in the negative sequence coordinates. Thus, only the high-order harmonic components other than the dc component need to be filtered out, which does not need to reselect the cut-off frequency of filters due to the change of motor operating frequency during ISF.

Traditional researches generally employ low pass filter or notch filter to filter out high-order harmonic components and preserve dc components. However, when the operating frequency changes, the parameters of the filter need to be adjusted accordingly, resulting in the detection error. To solve this, the paper presents a method based on angle integral filter to accurately measure the dc components of negative sequence current residuals, which can be incorporated into the original measurement algorithm. The principle of angular integral filtering can be expressed as:

$$\begin{bmatrix} \Delta i_{d\_dc}^n \\ \Delta i_{q\_dc}^n \end{bmatrix} = \frac{1}{\theta_n} \int_{\theta_e - \theta_n}^{\theta_e} \begin{bmatrix} \Delta i_d^n(\theta_e) \\ \Delta i_q^n(\theta_e) \end{bmatrix} d\theta_e \tag{27}$$

where θ_n is a complete electrical angle cycle, generally 2π. Therefore, the above fault index for ISF can be redefined as:

$$FI^n = \frac{3\sqrt{(\Delta i_{d\_dc}^n)^2 + (\Delta i_{q\_dc}^n)^2}}{\sqrt{2}\omega_e} \tag{28}$$

4.2. Fault Location

When the ISF occurs in PMSGs, the three-phase current will no longer maintain phase symmetry, and the phase angle difference between non-fault phases will decrease with the severity of fault. Figure 4 shows the phase angle change with ISF in phase *a*.

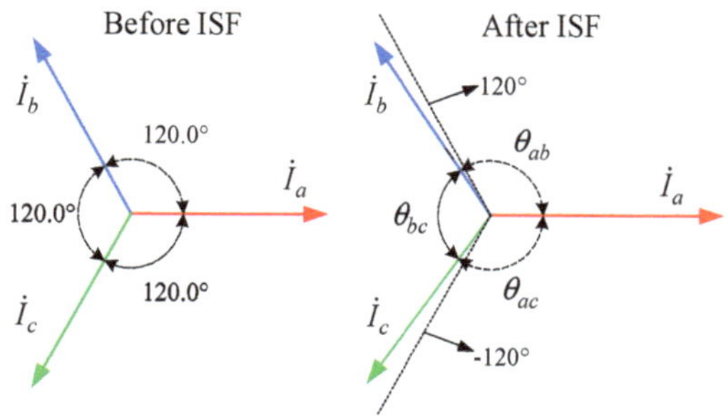

Figure 4. Phase angle changes with ISF in phase *a*.

Under the premise of ignoring the influence of higher harmonics, the stator three-phase current of PMSG can be expressed as:

$$\begin{cases} i_a = I_a \sin(\theta_e + \theta_a) \\ i_b = I_b \sin(\theta_e + \theta_b) \\ i_c = I_c \sin(\theta_e + \theta_c) \end{cases} \tag{29}$$

where I_j and θ_j ($j = a, b, c$) are the amplitude and initial phase of phase current, respectively. The phase angle differences between phase currents can be defined as:

$$\begin{cases} \theta_{ij} = |\theta_i - \theta_j|,\ |\theta_i - \theta_j| \le \pi \\ \theta_{ij} = 2\pi - |\theta_i - \theta_j|,\ \pi < |\theta_i - \theta_j| \le 2\pi \\ i, j = a, b, c;\ i \ne j \end{cases} \tag{30}$$

When the PMSG operates normally, there is $\theta_{ab} = \theta_{bc} = \theta_{ac} = \pi/3$. While the phase angle differences between the fault phase and the non-fault phases will increase ($>\pi/3$), and the phase angle difference between the non-fault phases will decrease ($<\pi/3$) when ISF occurs. Based on the above signature, the fault phase location indexes can be defined as follows:

$$\begin{cases} k_a = \frac{\theta_{ac} + \theta_{ab}}{2\theta_{bc}} \\ k_b = \frac{\theta_{ab} + \theta_{bc}}{2\theta_{ac}} \\ k_c = \frac{\theta_{ac} + \theta_{bc}}{2\theta_{ab}} \end{cases} \tag{31}$$

The theoretical value of fault location indexes when ISF occurs in different phases is shown in Table 1. There is a clear boundary between the location indexes of the fault phase and non-fault phases, which can be used to locate the fault phase. Under the same operating condition, the degree of deviation of the location indexes from 1 can reflect the severity of the fault to some extent.

Table 1. Theoretical value of fault location indexes with ISF.

Fault Phase	k_a	k_b	k_c
a	>1	<1	<1
b	<1	>1	<1
c	<1	<1	>1

For phase angle measurement of three-phase current, the currents can be projected to a rotating coordinate system, which can be expressed as:

$$\begin{cases} i_j^d = i_j \cos\theta_e = \frac{1}{2} I_j \sin(2\theta_e + \theta_j) - \frac{1}{2} I_j \sin\theta_j \\ \quad = i_{j\_2nd}^d \sin(2\theta_e + \theta_j) + i_{j\_dc}^d \\ i_j^q = i_j \sin\theta_e = -\frac{1}{2} I_j \cos(2\theta_e + \theta_j) + \frac{1}{2} I_j \cos\theta_j \\ \quad = i_{j\_2nd}^q \cos(2\theta_e + \theta_j) + i_{j\_dc}^q \end{cases} \quad (32)$$

where $j = a, b, c$. Subscripts '*2nd*' and '*dc*' are the second harmonic and dc component respectively. According to Equation (32), the dc component in the rotating coordinate systems contains the initial phase angle of the three-phase current. Therefore, the angular integral filtering method mentioned above can be used to extract the dc components first, and then the reconstruction can be carried out as follows:

$$\theta_j = \tan^{-1}\left(-\frac{i_{j\_dc}^d}{i_{j\_dc}^q}\right) \quad (33)$$

4.3. Overall Process of Fault Detection and Location

The overall process of ISF detection and location in PMSG in this paper is shown in Figure 5. Firstly, the residual currents on *dq*-axis in the ISF state is observed by the Luenberger observer. Secondly, the dc components in the negative sequence current residuals are extracted by negative sequence Park transformation and angular integral filter to construct fault detection index, and the initial phase angle of three-phase currents are calculated by the method of rotating coordinate projection, on which the fault location indexes are constructed. Finally, the measured parameters are compared with the normal conditions to determine whether the ISF occurs in PMSG, thus send out alarm signal or trip the circuit breaker according to the severity of the fault.

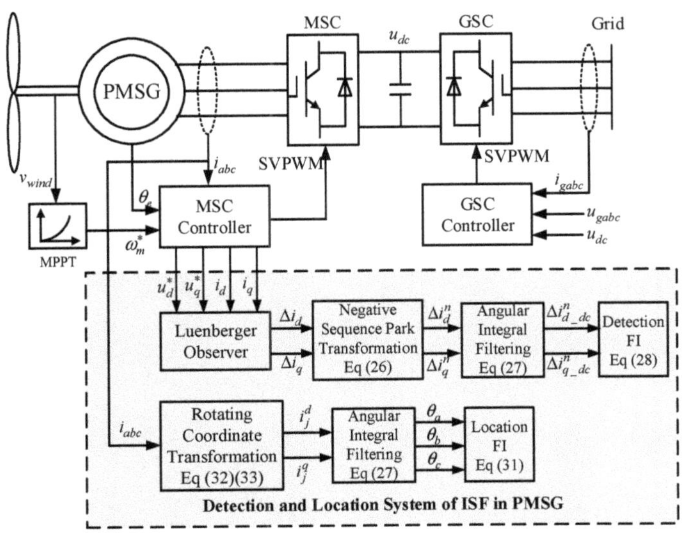

Figure 5. Overall process of ISF detection and location in PMSG.

5. Simulation Verification

In order to verify the effectiveness and rationality of the ISF detection and location strategy proposed in this paper, the ISF model of PMSG is built in Matlab/Simulink for simulation testing. The parameters of PMSG used in the simulation are shown in Table 2. Before the simulation time t = 1.3 s, the PMSG is in a healthy state. When t = 1.3 s, the ISF occurs in phase a by closing the breaker of the short circuit. The short circuit parameters are set to μ = 0.15, R_f = 0.002.

Table 2. Parameters of the PMSG based wind power system.

Specification	Symbol	Value
Blade radius	R	38.8 m
Air density	ρ	1.225 kg/m^3
Optimal Power Coefficient	C_{pmax}	0.48
Rated wind speed	v_r	12 m/s
Rated power	P_r	2.4 MW
Phase resistance	R_s	0.01 Ω
d-axis inductance	L_d	5 mH
q-axis inductance	L_q	5 mH
Rotor inertia	J	16,000 kg·m^2
PM flux linkage	λ_{PM}	10 Wb
Pole-pair numbers	n_p	40

5.1. ISF Detection under Steady Wind Speed

The simulated wind speed v_{wind} is set as 10 m/s. Figure 6 shows the simulation waveforms of three-phase current, rotation speed and electromagnetic torque of the PMSG in healthy and failure states. The three-phase current will lose symmetry after ISF, and harmonic components appear in the speed and electromagnetic torque, which conforms to the basic characteristics of the fault.

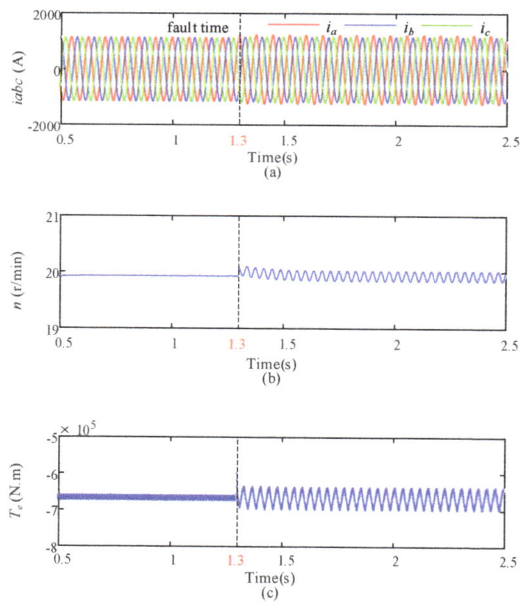

Figure 6. Simulation results. (**a**) Three-phase current of stator; (**b**) speed; (**c**) electromagnetic torque.

The waveforms of current residuals on *dq*-axis are shown in Figure 7a–c is the spectrum of current residuals after the fault. Under normal conditions, the current residuals are almost equal to zero, so the spectrum analysis of the signal before the fault is not very meaningful. After the fault, obvious dc component and second harmonics appear in the residuals. The second harmonic on *d*-axis and *q*-axis have almost the same amplitude, while the difference between dc components is mainly related to the fault location, which is consistent with the theoretical analysis result of Equation (23) above.

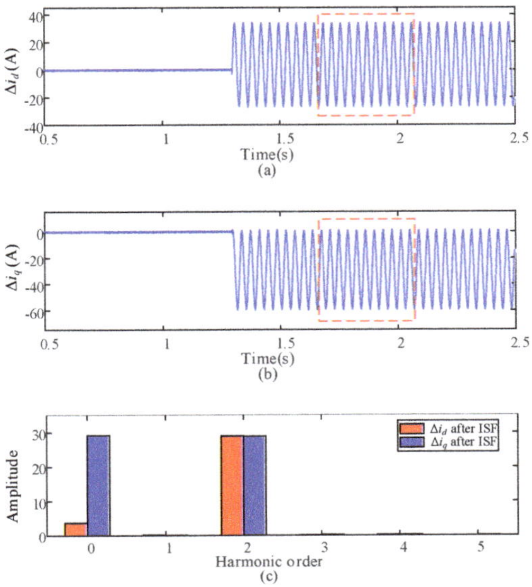

Figure 7. Simulation results of current residuals. (**a**) *d*-axis current residual; (**b**) *q*-axis current residual; (**c**) spectrum of current residuals after ISF.

Figure 8 shows the simulation results of fault detection and phase location indexes under health and fault conditions. When the PMSG is in a healthy state, the fault detection index is close to 0, and the three-phase location indexes are all close to 1. After the ISF occurs, the detection index mutates and stabilizes to 1.044 within 0.05 s. The location indexes show a difference after 0.1 s, $k_a > 1$, and it is significantly greater than indexes of other two phases, which corresponds to the occurrence of ISF in phase *a*.

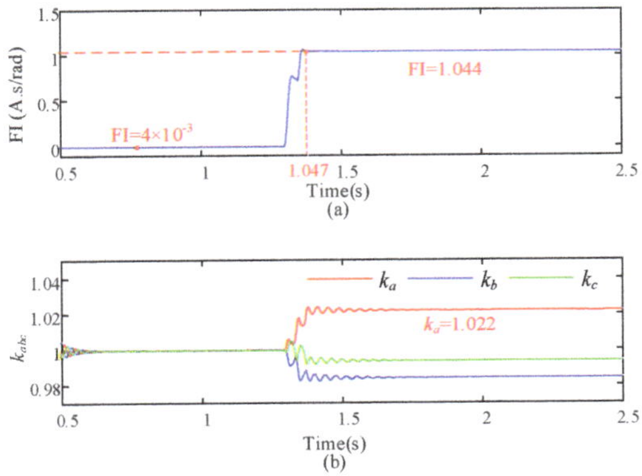

Figure 8. Simulation results of fault detection and fault location indexes. (**a**) Fault detection index; (**b**) fault phase location indexes.

Considering that the interturn fault are not handled in time, which may lead to further failures, it is necessary to study the universality of fault indexes in continuous state. The degree of failure is mainly determined by the shorted turn ratio μ and fault resistance R_f. Figure 9 shows the influence of different μ and R_f on fault indexes under continuous operation state. In Figure 9a,b, the fault resistance R_f is set to 0.02, the ratio μ increases by 0.05 every 0.5 s after 1 s. With the increase of shorted turn ratio μ, both detection index *FI* and fault phase location indexes k_{abc} increase proportionally. In Figure 9c,d, the shorted turn ratio μ is set to 0.15, the fault resistance R_f rises from 0.005 to 0.04. As the fault resistance R_f increases, the detection index *FI* and fault phase location indexes k_{abc} decrease inversely. When $R_f > 0.03$, k_{abc} has almost become stable with little change. Therefore, the proposed fault detection index and fault phase location indexes can reflect the severity of the ISF.

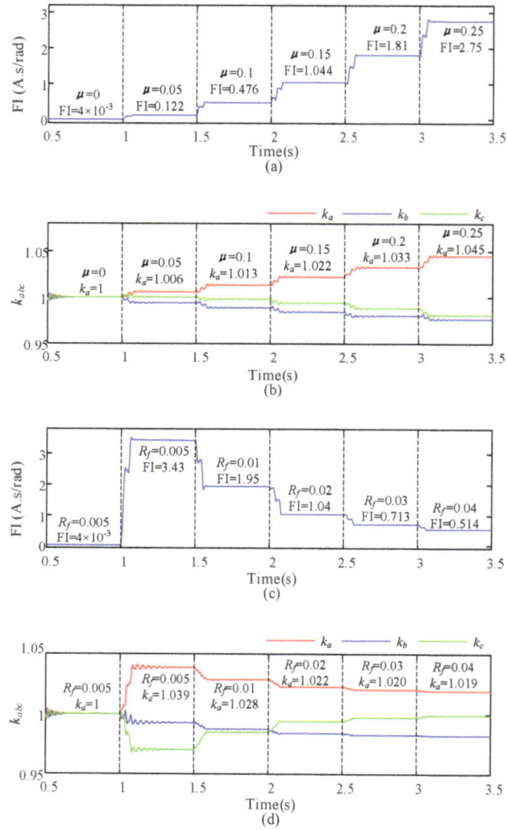

Figure 9. Fault detection index and fault phase location indexes under different fault severity. (**a**) Detection index with different μ when $R_f = 0.02$; (**b**) location indexes with different μ when $R_f = 0.02$; (**c**) detection index with different R_f when $\mu = 0.15$; (**d**) location indexes with different R_f when $\mu = 0.15$.

Figure 10 shows the changes of detection index and location index under different wind speeds. Since the wind generator work under MPPT mode, different wind speeds mean that the PMSG will track the maximum power point at different optimal speeds. In Figure 10a,c, the changes of wind speed has little impact on the detection index *FI* under the same fault severity. Therefore, the threshold of fault detection index is almost universal for different working wind speeds. In Figure 10b,d, the location index of fault phase k_a decreases with the increase of wind speed, but the decrease is not large, and is obviously larger than that of healthy state, so the indexes are still effective in the location of the ISF.

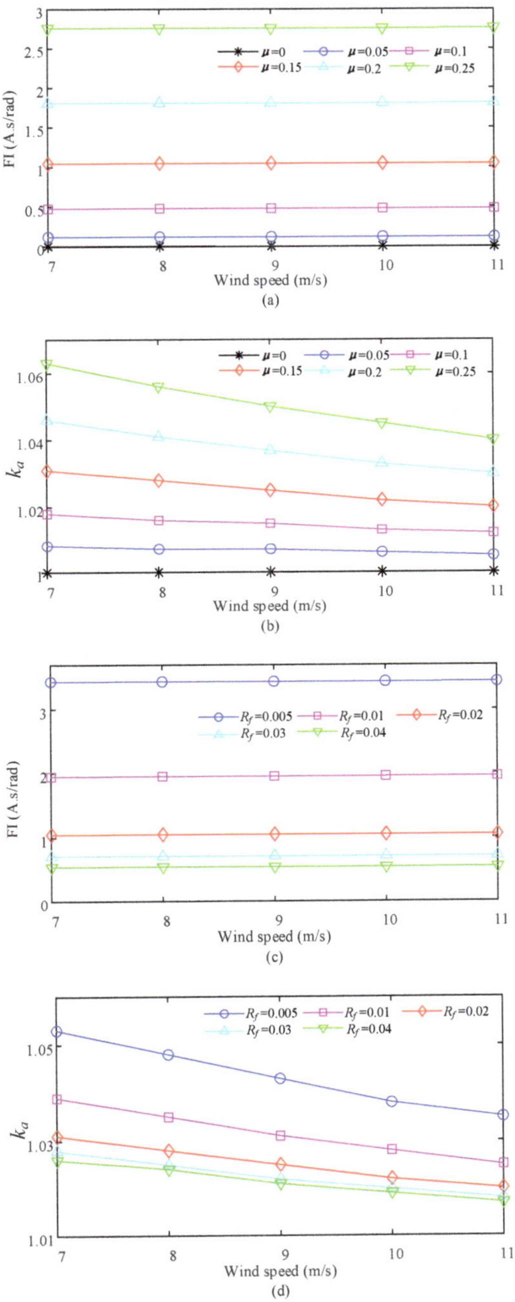

Figure 10. Fault detection index and fault phase location index under different wind speeds. (**a**) Detection index *FI* with different μ; (**b**) location index k_a with different μ; (**c**) detection index *FI* with different R_f; (**d**) location index k_a with different R_f.

5.2. ISF Detection under Variable Wind Speed

In the above, the fault detection is carried out under the steady state of a specific wind speed, while the PMSG mostly works under continuously changing wind speed, so the following will discuss the ISF detection under variable wind speed. Figure 11a shows the stepped wind speed input. The optimum, and actual speed, of PMSG are shown in Figure 11b. Under MPPT control mode, the speed of PMSG can still track the basic trajectory of the optimum speed, but there is obvious fluctuation caused by ISF. In Figure 11c, the fault detection index of the same fault severity is almost the same at different wind speeds. At the moment of step change of wind speed, the index value will fluctuate, but the fluctuation is limited, which needs to be considered when selecting the fault threshold. However, the fault phase location index may change suddenly when the wind speed surges, and the indexes of non-fault phases may be greater than 1, thus affecting the location results. Therefore, the fault phase location can be carried out after the fault detection index is stable.

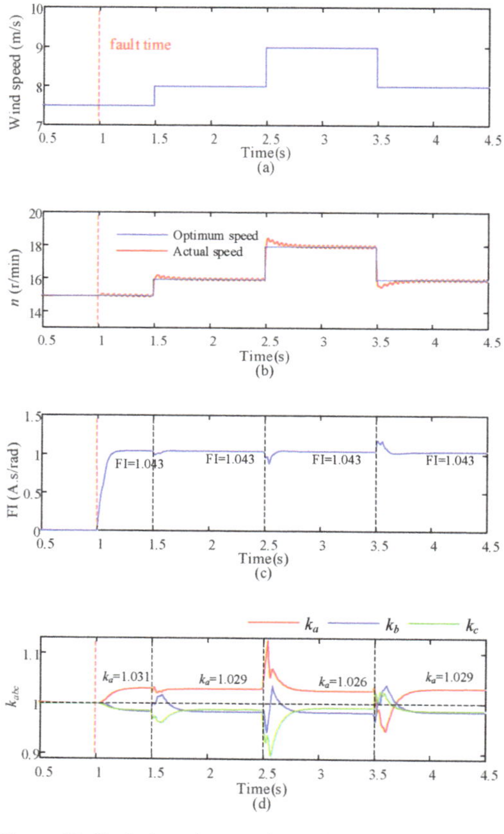

Figure 11. Fault detection results under stepped wind speed. (**a**) Stepped wind speed; (**b**) pptimum and actual speeds of PMSG; (**c**) fault detection index; (**d**) fault phase location indexes.

Under the stepped wind speed in Figure 11a, the fault detection index curves of different fault severity are shown in Figure 12. From the simulation results, the sudden change of the wind speed will cause the fluctuation of detection indexes under different fault severity, but the overall trend is stable, which will not have a great impact on the final fault diagnosis.

On the basis of above simulation under stepped wind speed, when the wind generator operates at random wind speed as shown in Figure 13a, the simulation results are shown in Figure 13. The speed and electromagnetic torque of the PMSG have obvious harmonics after ISF. Under the continuously changing wind speed, the detection index cannot reach a relatively stable state due to the measurement error caused by the low frequency of the

PMSG, but it is obviously larger than the index curve under the healthy state, so it is still effective for ISF diagnosis.

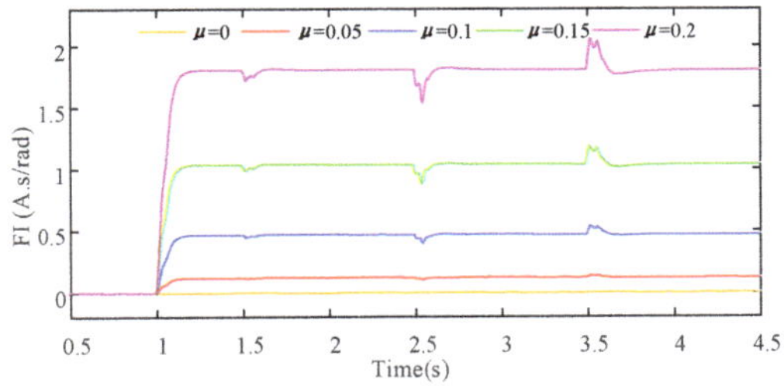

Figure 12. Fault detection results under stepped wind speed with different μ.

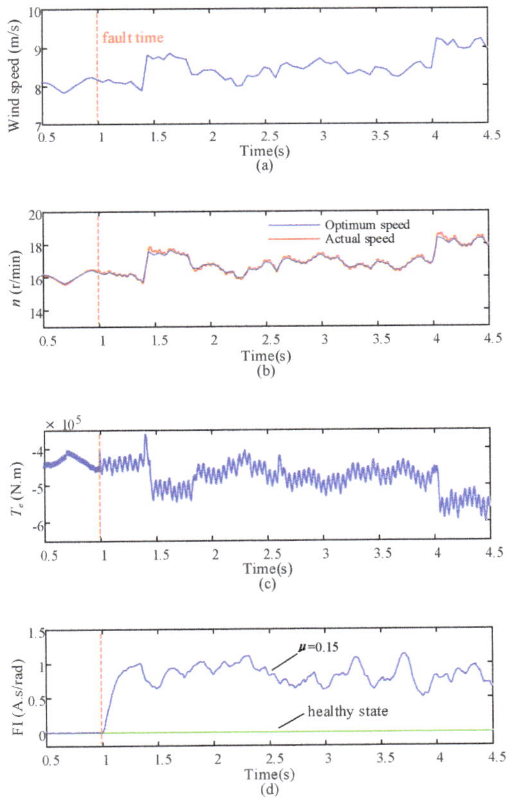

Figure 13. Fault detection results at random wind speed. (**a**) Random wind speed; (**b**) optimum and actual speeds of PMSG; (**c**) electromagnetic torque; (**d**) fault detection index.

Under the random wind speed in Figure 13a, the fault detection index curves of different fault severity are shown in Figure 14. Compared with stepped wind speed, the fluctuation of detection index is more intense with the deepening of fault severity, but there are still clear boundaries between different curves. If subsequent fault severity prediction is required after diagnosis, it can be distinguished by calculating the energy spectrum or power spectrum of the curves.

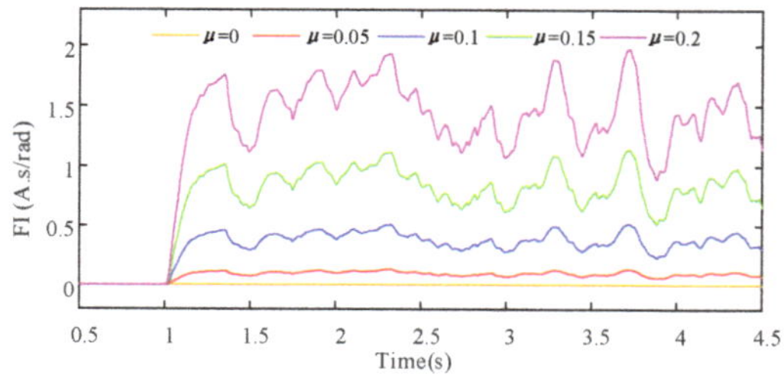

Figure 14. Fault detection results at random wind speed with different μ.

6. Conclusions

A robust fault detection and location strategy for ISF in PMSGs based on negative sequence current residuals is presented in this paper. The proposed method utilizes the second harmonics in the residuals as a reliable fault signature rather than the dc components since the dc components are not only related to the severity of the ISF, but also affected by the short circuit location and errors of the motor parameters. The current residuals on *dq*-axis are acquired by the Luenberger observer based on the expanded mathematical model of PMSG. The second harmonics are extracted through negative sequence park transform and angular integral filtering equivalently. By reconstructing the negative sequence current residuals, the defined fault detection index has great robustness to the normal speed fluctuation of wind generator. In addition, the unbalance characteristics of three-phase current after ISF can reflect the location of the fault phase, which is the basis for the definition of location indexes. However, there are mutations in location indexes, so the location of fault phase can be done after detection index is stable. The extensive simulation results in Matlab/Simulink validated the effectiveness and stability of fault detection and fault phase location method proposed in this paper.

Author Contributions: Conceptualization, T.W. and S.C.; methodology, W.D. and X.L. (Xindong Li); software, W.D.; validation, Y.Z., X.L. (Xiaobao Liu) and X.L. (Xindong Li).; formal analysis, S.C.; investigation, T.W.; resources, T.W.; data curation, T.W. and W.D.; writing—original draft preparation, S.C.; writing—review and editing, S.C.; visualization, X.L. (Xiaobao Liu); supervision, T.W.; project administration, T.W.; funding acquisition, T.W. All authors have read and agreed to the published version of the manuscript.

Funding: This work was supported by State Key Laboratory of Smart Grid Protection and Control, Nari Group Corporation, Nanjing, Jiangsu, 211106, China. (Project No. SGNR0000KJJS2200310).

Data Availability Statement: Not applicable.

Conflicts of Interest: The authors declare no conflict of interest.

References

1. Lee, H.; Jeong, H.; Kim, S.W. Diagnosis of Interturn Short-Circuit Fault in PMSM by Residual Voltage Analysis. In Proceedings of the 2018 International Symposium on Power Electronics, Electrical Drives, Automation and Motion (SPEEDAM), Amalfi, Italy, 20–22 June 2018; pp. 160–164.
2. Moosavi, S.S.; Djerdir, A.; Amirat, Y.; Khaburi, D.A. Demagnetization fault diagnosis in permanent magnet synchronous motors: A review of the state-of-the-art. *J. Magn. Magn. Mater.* **2015**, *391*, 203–212. [CrossRef]
3. Yao, J.; Pei, J.; Xu, D.; Liu, R.; Wang, X.; Wang, C.; Li, Y. Coordinated control of a hybrid wind farm with DFIG-based and PMSG-based wind power generation systems under asymmetrical grid faults. *Renew. Energy* **2018**, *127*, 613–629. [CrossRef]
4. Yao, J.; Liu, R.; Zhou, T.; Hu, W.; Chen, Z. Coordinated control strategy for hybrid wind farms with DFIG-based and PMSG-based wind farms during network unbalance. *Renew. Energy* **2017**, *105*, 748–763. [CrossRef]
5. Yaramasu, V.; Wu, B.; Alepuz, S.; Kouro, S. Predictive Control for Low-Voltage Ride-Through Enhancement of Three-Level-Boost and NPC-Converter-Based PMSG Wind Turbine. *IEEE Trans. Ind. Electron.* **2014**, *61*, 6832–6843. [CrossRef]
6. Zafarani, M.; Bostanci, E.; Qi, Y.; Goktas, T.; Akin, B. Interturn Short-Circuit Faults in Permanent Magnet Synchronous Machines: An Extended Review and Comprehensive Analysis. *IEEE J. Emerg. Sel. Top. Power Electron.* **2018**, *6*, 2173–2191. [CrossRef]
7. Cintron-Rivera, J.G.; Foster, S.N.; Strangas, E.G. Mitigation of turn-to-turn faults in fault tolerant permanent magnet synchronous motors. *IEEE Trans. Energy Convers.* **2015**, *30*, 465–475. [CrossRef]

8. Gandhi, A.; Corrigan, T.; Parsa, L. Recent Advances in Modeling and Online Detection of Stator Interturn Faults in Electrical Motors. *IEEE Trans. Ind. Electron.* **2011**, *58*, 1564–1575. [CrossRef]
9. Moon, S.; Lee, J.; Jeong, H.; Kim, S.W. Demagnetization Fault Diagnosis of a PMSM Based on Structure Analysis of Motor Inductance. *IEEE Trans. Ind. Electron.* **2016**, *63*, 3795–3803. [CrossRef]
10. Chen, Y.; Liang, S.; Li, W.; Liang, H.; Wang, C. Faults and Diagnosis Methods of Permanent Magnet Synchronous Motors: A Review. *Appl. Sci.* **2019**, *9*, 2116. [CrossRef]
11. Bhuiyan, E.A.; Akhand, M.M.A.; Das, S.K.; AliMd, F.; Tasneem, Z.; Islam, Md.R.; Saha, D.K.; Badal, F.R.; Ahamed, Md.H.; Moyeen, S.I. A Survey on Fault Diagnosis and Fault Tolerant Methodologies for Permanent Magnet Synchronous Machines. *Int. J. Autom. Comput.* **2020**, *17*, 763–787. [CrossRef]
12. Qian, H.; Guo, H.; Ding, X. Modeling and Analysis of Interturn Short Fault in Permanent Magnet Synchronous Motors With Multistrands Windings. *IEEE Trans. Power Electron.* **2016**, *31*, 2496–2509. [CrossRef]
13. Ben Khader Bouzid, M.; Champenois, G.; Maalaoui, A.; Tnani, S. Efficient Simplified Physical Faulty Model of a Permanent Magnet Synchronous Generator Dedicated to the Stator Fault Diagnosis Part I: Faulty Model Conception. *IEEE Trans. Ind. Appl.* **2017**, *53*, 2752–2761. [CrossRef]
14. Hang, J.; Zhang, J.; Xia, M.; Ding, S.; Hua, W. Interturn Fault Diagnosis for Model-Predictive-Controlled-PMSM Based on Cost Function and Wavelet Transform. *IEEE Trans. Power Electron.* **2020**, *35*, 6405–6418. [CrossRef]
15. Yang, Z.; Wang, Y.; Lv, J. Survey of modern Fault Diagnosis methods in networks. In Proceedings of the 2012 International Conference on Systems and Informatics (ICSAI2012), Yantai, China, 19–20 May 2012; pp. 1640–1643.
16. Fu, X. Statistical machine learning model for capacitor planning considering uncertainties in photovoltaic power. *Prot. Control Mod. Power Syst.* **2022**, *7*, 51–63. [CrossRef]
17. Tang, S.; Yuan, S.; Zhu, Y. Deep Learning-Based Intelligent Fault Diagnosis Methods Toward Rotating Machinery. *IEEE Access* **2020**, *8*, 9335–9346. [CrossRef]
18. Skowron, M.; Orlowska-Kowalska, T.; Kowalski, C.T. Detection of Permanent Magnet Damage of PMSM Drive Based on Direct Analysis of the Stator Phase Currents Using Convolutional Neural Network. *IEEE Trans. Ind. Electron.* **2022**, *69*, 13665–13675. [CrossRef]
19. Yan, H.; Xu, Y.; Cai, F.; Zhang, H.; Zhao, W.; Gerada, C. PWM-VSI Fault Diagnosis for a PMSM Drive Based on the Fuzzy Logic Approach. *IEEE Trans. Power Electron.* **2019**, *34*, 759–768. [CrossRef]
20. Delgado, M.; García, A.; Ortega, J.A.; Cárdenas, J.J.; Romeral, L. Multidimensional intelligent diagnosis system based on Support Vector Machine Classifier. In Proceedings of the 2011 IEEE International Symposium on Industrial Electronics, Gdansk, Poland, 27–30 June 2011; pp. 2124–2131.
21. Haje Obeid, N.; Battiston, A.; Boileau, T.; Nahid-Mobarakeh, B. Early Intermittent Interturn Fault Detection and Localization for a Permanent Magnet Synchronous Motor of Electrical Vehicles Using Wavelet Transform. *IEEE Trans. Transp. Electrif.* **2017**, *3*, 694–702. [CrossRef]
22. Wang, B.; Wang, J.; Griffo, A.; Sen, B. Stator Turn Fault Detection by Second Harmonic in Instantaneous Power for a Triple-Redundant Fault-Tolerant PM Drive. *IEEE Trans. Ind. Electron.* **2018**, *65*, 7279–7289. [CrossRef]
23. Hu, R.; Wang, J.; Mills, A.R.; Chong, E.; Sun, Z. High-Frequency Voltage Injection Based Stator Interturn Fault Detection in Permanent Magnet Machines. *IEEE Trans. Power Electron.* **2021**, *36*, 785–794. [CrossRef]
24. Heydarzadeh, M.; Zafarani, M.; Akin, B.; Nourani, M. Automatic fault diagnosis in PMSM using adaptive filtering and wavelet transform. In Proceedings of the 2017 IEEE International Electric Machines and Drives Conference (IEMDC), Miami, FL, USA, 21–24 May 2017; pp. 1–7.
25. Wu, Y.; Zhang, Z.; Li, Y.; Sun, Q. Open-Circuit Fault Diagnosis of Six-Phase Permanent Magnet Synchronous Motor Drive System Based on Empirical Mode Decomposition Energy Entropy. *IEEE Access* **2021**, *9*, 91137–91147. [CrossRef]
26. He, S.; Shen, X.; Jiang, Z. Detection and Location of Stator Winding Interturn Fault at Different Slots of DFIG. *IEEE Access* **2019**, *7*, 89342–89353. [CrossRef]
27. Yin, Z.; Sui, Y.; Zheng, P.; Yang, S.; Zheng, Z.; Huang, J. Short-Circuit Fault-Tolerant Control Without Constraint on the D-Axis Armature Magnetomotive Force for Five-Phase PMSM. *IEEE Trans. Ind. Electron.* **2022**, *69*, 4472–4483. [CrossRef]
28. Moon, S.; Jeong, H.; Lee, H.; Kim, S.W. Interturn Short Fault Diagnosis in a PMSM by Voltage and Current Residual Analysis With the Faulty Winding Model. *IEEE Trans. Energy Convers.* **2018**, *33*, 190–198. [CrossRef]
29. Yin, X.X.; Lin, Y.G.; Li, W.; Liu, H.W.; Gu, Y.J. Fuzzy-Logic Sliding-Mode Control Strategy for Extracting Maximum Wind Power. *IEEE Trans. Energy Convers.* **2015**, *30*, 1267–1278. [CrossRef]
30. Romeral, L.; Urresty, J.C.; Riba Ruiz, J.R.; Garcia Espinosa, A. Modeling of Surface-Mounted Permanent Magnet Synchronous Motors with Stator Winding Interturn Faults. *IEEE Trans. Ind. Electron.* **2011**, *5*, 1576–1585. [CrossRef]
31. Otava, L.; Buchta, L. PMSM stator winding faults modelling and measurement. In Proceedings of the 2015 7th International Congress on Ultra Modern Telecommunications and Control Systems and Workshops (ICUMT), Brno, Czech Republic, 6–8 October 2015; pp. 138–143.
32. Andersson, A.; Thiringer, T. Motion Sensorless IPMSM Control Using Linear Moving Horizon Estimation with Luenberger Observer State Feedback. *IEEE Trans. Transp. Electrif.* **2018**, *4*, 464–473. [CrossRef]
33. Gurusamy, V.; Bostanci, E.; Li, C.; Qi, Y.; Akin, B. A Stray Magnetic Flux-Based Robust Diagnosis Method for Detection and Location of Interturn Short Circuit Fault in PMSM. *IEEE Trans. Instrum. Meas.* **2021**, *70*, 1–11. [CrossRef]

energies

MDPI

Article

Influence of Stator Core Seams on No-Load Performance of Module-Combined Stator Permanent Magnet Motor and Its Weakening Method

Shaonan Sun, Guihong Feng *, Yan Li and Bingyi Zhang

School of Electrical Engineering, Shenyang University of Technology, Shenyang 110870, China; mytcdssn@163.com (S.S.)
* Correspondence: fenggh@sut.edu.cn; Tel.: +86-13591690366

Abstract: In this paper, a module-combined stator is proposed, which is used for large and ultra-low-speed permanent magnet synchronous motors, and the influence of stator core seams on the no-load performance is studied. A method is proposed to weaken the negative influence of stator iron core seams on the no-load performance of permanent magnet synchronous motors. Firstly, the magnetic circuit model of the motor considering the stator iron core seams was deduced theoretically, and the selection principle of stator core seam number was given a description. The influence of different seam parameters on the no-load performance and the influence of different pole-slot fits and the number of parallel branches on the no-load performance are analyzed. The proposed structure of the stator iron core, which can weaken the influence of stator iron core seams on the no-load performance of the motor, is proposed. Using analysis and simulation experiments, the effectiveness of the proposed stator iron core structures in weakening the negative influence of the stator iron core seams on the no-load performance was verified.

Keywords: module-combine stator; PMSM (permanent magnet synchronous motors); stator core seams; no-load performance; auxiliary seams

Citation: Sun, S.; Feng, G.; Li, Y.; Zhang, B. Influence of Stator Core Seams on No-Load Performance of Module-Combined Stator Permanent Magnet Motor and Its Weakening Method. *Energies* **2023**, *16*, 4126. https://doi.org/10.3390/en16104126

Academic Editor: Gianluca Brando

Received: 29 March 2023
Revised: 25 April 2023
Accepted: 13 May 2023
Published: 16 May 2023

1. Introduction

For large and ultra-low-speed motors, module-combined stator permanent magnet motors are used to reduce the assembly difficulty and improve the fault tolerance [1–3]. The stator module combination structure of permanent magnet motors has three main schemes:

Scheme 1: A single set of stator teeth and a stator yoke are superposed to form a tooth–yoke sub-module, and the coil is wound on the tooth–yoke sub-module [4,5];

Scheme 2: A stator sector piece with a single set of stator teeth is separately superposed onto a sub-module stator, and the wound coil is directly wound onto the stator tooth module. This stator tooth module and the stator yoke are fixed via a separate connecting piece or a pigeon tail slot on the stator sheet [6,7];

Scheme 3: A stator sector piece that can wind one group of three-phase windings is separately superposed into a sub-module stator. The windings are wound on a single tooth when the seams of the stator core module are in the yoke and the pitch $y = 1$, while the seams of the stator core module are in the tooth or pitch $Y > 1$. The windings are divided into two different span specifications: large and small-span windings [8–12].

A novel cogging torque mitigation method for modular permanent magnet (PM) machines with flux gaps in alternate stator teeth has been proposed [13,14]. The influence of these flux gaps on the electromagnetic performance of modular PM machines, such as on the winding factor, open-circuit air-gap flux densities, back electromotive force, cogging torque, on-load torque, inductance, magnetic saturation and copper losses, were comprehensively investigated and general rules have been established [15].

In this paper, a module-combined stator is proposed for large and ultra-low-speed permanent magnet synchronous motors, and the influence of the stator core seams was studied. For large and ultra-low-speed motors (<60 rpm), copper loss (>80% total loss) is mainly dependent on the no-load back electromotive force and the same coil parameters and cogging torque can significantly influence the motor parameter measurement accuracy of the converter; therefore, this paper mainly studied the performance of the no-load back electromotive force and cogging torque. Firstly, a magnetic circuit model considering the stator core seams was established. The effect of the dimension parameters of the seams on the no-load performance of the motor was studied using different pole slot fits and number of parallel branches. Based on the comparison of the effects of different parts of the stator core seams, two kinds of stator sheet structures are proposed to weaken the adverse effects of the stator core seams.

2. Magnetic Circuit Model with Consideration of Stator Core Seams

A schematic of stator core seams with an outer rotor permanent magnet surface mounting motor is given in Figure 1. In the enlarged view around the stator core seams in Figure 1b, N_{seam} is the seam number; h_{j2} is the height of the rotor yoke; h_m and r_2 are the inner diameter and thickness of the permanent magnet, respectively; g_0 is the air gap between the stator and rotor; h_{01} and b_{01} are the height and width of the stator notch, respectively; b_t is the regular tooth width; b_{t11} and b_{t12} are the left tooth width and the right tooth width, respectively, with the seam on the tooth; b_{seam} is the seam width; r_1 is the radius of the slot bottom; h_{j1} is the height of the stator yoke; and h_{s1} is the radial projection height from the slot shoulder bottom to the slot notch.

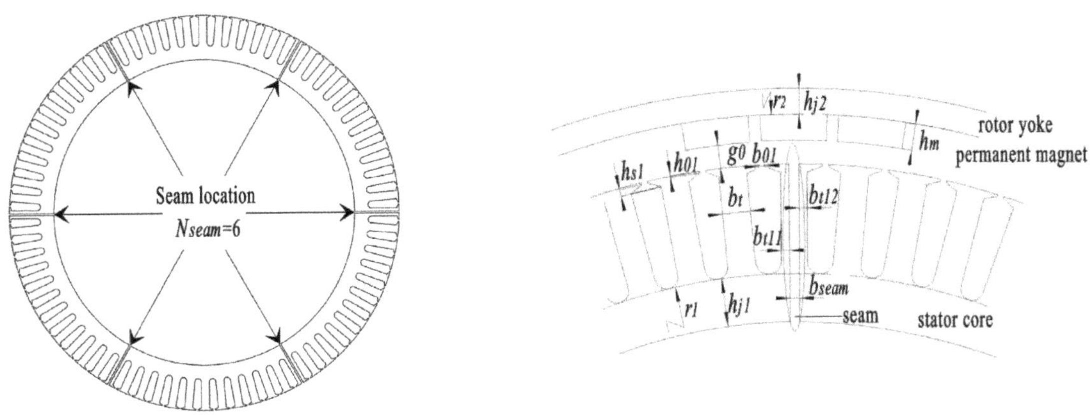

(**a**)The stator core with 6 seams (**b**) Enlarged view around the stator core seam

Figure 1. The schematic drawing of stator with seams (N_{seam} = 6).

2.1. The Selection Principle of Seam Number

The number of slots per pole per phase in a multi-pole, ultra-low-speed permanent magnet synchronous motor with fewer slots is shown in Expression (1).

$$q = \frac{Q_s}{2pm} = b + \frac{c}{d} \tag{1}$$

where q is the number of slots per pole per phase, Q_S is the number of stator slots, p is the number of motor pole pairs, m is motor phase, and c and d are the minimum forms of the numerator or denominator.

The maximum number of pole phase groups is $2p/d$. When d is even, the value of seam number N_{seam} should be the divisor of $2p/d$ because the large span coil must connect at least one pole phase group as shown in Figure 2a. When d is odd, the value of seam number N_{seam} should be the divisor of p/d because the large span coil must connect at least two pole phase groups as shown in Figure 2b.

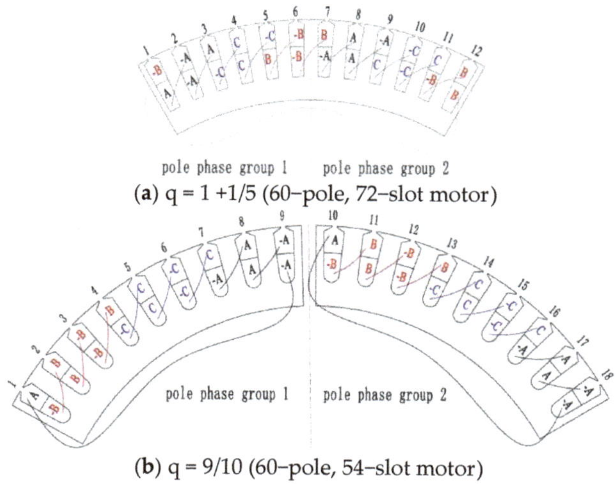

Figure 2. Pole phase groups with two different pole slot combinations.

2.2. The Magnetic Circuit

The magnetic circuit model considering the stator core seams is given in Figure 3. In order to facilitate the modeling, the following assumptions were made:

1. The flux in the left tooth and right tooth with the seam on the tooth are linearly proportional to the tooth width;
2. The magnetic resistance of each segment was simplified to a rectangle in the calculation of the magnetic resistance;
3. The magnetic permeability of the ferromagnetic material was set to be invariable, and the corresponding magnetic resistance was assumed to be invariable.

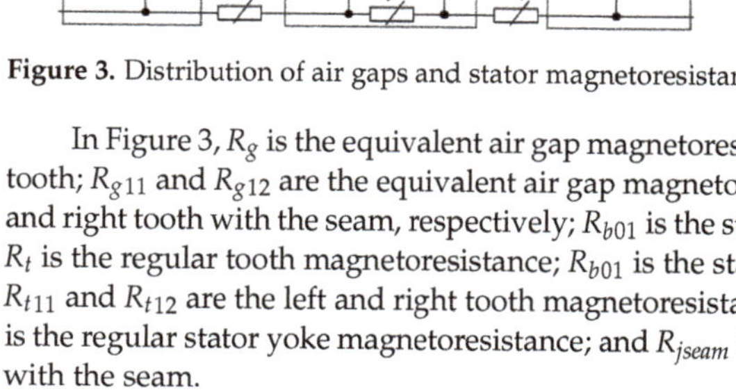

Figure 3. Distribution of air gaps and stator magnetoresistance around the stator core seam.

In Figure 3, R_g is the equivalent air gap magnetoresistance corresponding to the regular tooth; R_{g11} and R_{g12} are the equivalent air gap magnetoresistances corresponding to the left and right tooth with the seam, respectively; R_{b01} is the stator slot leakage magnetoresistance; R_t is the regular tooth magnetoresistance; R_{b01} is the stator slot leakage magnetoresistance; R_{t11} and R_{t12} are the left and right tooth magnetoresistances with the seam, respectively; R_j is the regular stator yoke magnetoresistance; and R_{jseam} is the stator yoke magnetoresistance with the seam.

The expression of magnetoresistance R is shown in (2), where l is the length of the magnetic path, μ is the magnetoconductivity of the medium, and S is the sectional area of the medium.

$$R = \frac{l}{\mu S} \tag{2}$$

Based on the expression (2), R_g, R_{g11}, and R_{g12} can be expressed as:

$$\begin{cases} R_g = \dfrac{Q_s k_\delta g_0}{2\pi\mu_0(r_2 - h_m - 0.5g_0)L_{Fe}} \\ R_{g11} = \dfrac{Q_s k_{\delta 1} \frac{b_{t11} + b_{t12}}{b_{t11}} g_0}{2\pi\mu_0(r_2 - h_m - 0.5g_0)L_{Fe}} \\ R_{g12} = \dfrac{Q_s k_{\delta 1} \frac{b_{t11} + b_{t12}}{b_{t12}} g_0}{2\pi\mu_0(r_2 - h_m - 0.5g_0)L_{Fe}} \end{cases} \tag{3}$$

where k_δ and $k_{\delta 1}$ are the air gap coefficients corresponding to the regular stator tooth and stator tooth with the seam, respectively; g_0 is the air gap between stator and rotor, μ_0 is the vacuum magnetic permeability, r_2 is the inner diameter of rotor yoke, h_m is the magnetization direction length of permanent magnets, L_{Fe} is the iron core length of the stator and rotor, b_{t11} is the tooth width on the left of the seam, and b_{t12} is the tooth width on the right of the seam. Referring to reference [16–18], k_δ and $k_{\delta 1}$ can be expressed as:

$$\begin{cases} k_\delta = \dfrac{1}{1 - \frac{Q_s b_{01}}{2\pi(r_2 - h_m - g_0)}\sigma} \\ \sigma = \frac{2}{\pi}\left\{\tan^{-1}\frac{0.5b_{01}}{h_m + g_0} - \frac{h_m + g_0}{b_{01}}\ln\left(1 + \left(\frac{0.5b_{01}}{h_m + g_0}\right)^2\right)\right\} \\ k_{\delta 1} = \dfrac{1}{1 - \frac{Q_s(b_{01} + b_{seam})}{4\pi(r_2 - h_m - g_0)}\sigma_1} \\ \sigma_1 = \frac{2}{\pi}\ln\left\{\begin{array}{c} \tan^{-1}\frac{b_{01} + b_{seam}}{4(h_m + g_0)} - \\ \frac{2(h_m + g_0)}{b_{01} + b_{seam}}\left(1 + \left(\frac{b_{01} + b_{seam}}{4(h_m + g_0)}\right)^2\right) \end{array}\right\} \end{cases} \tag{4}$$

where b_{01} is the width of the top stator slot, and b_{seam} is the width of the stator core seam. The stator tooth magnetoresistance and yoke magnetoresistance can be expressed as follows, where μ_{Fe} is the relative magnetic permeability of the stator:

$$\begin{cases} R_t = \dfrac{r_2 - h_m - g_0 - r_1}{\mu_{Fe}\mu_0 b_t L_{Fe}} \\ R_{t11} = \dfrac{r_2 - h_m - g_0 - r_1}{\mu_{Fe}\mu_0 b_{t11} L_{Fe}} \\ R_{t12} = \dfrac{r_2 - h_m - g_0 - r_1}{\mu_{Fe}\mu_0 b_{t12} L_{Fe}} \\ R_{jseam} = \dfrac{b_{seam}}{\mu_0 h_{j1} L_{Fe}} \end{cases} \tag{5}$$

where r_1 is the outer diameter of the stator yoke, μ_{fe} is the magnetic permeability of the stator core, b_t is the width of the stator tooth without seam, and h_{j1} is the radial height of the stator yoke.

3. Influence of Seams on No-Load Performance

3.1. Influence of Different Seam Numbers and Seam Widths on No-Load Performance

The motor parameters are given in Table 1. The winding end connection diagram is shown in Figure 4 in the case of N_{seam} = 3, a = 3, a_{Unit} = 1 and N_{seam} = 6, a = 6, a_{Unit} = 1, where a is the number of parallel branches of the whole stator and a_{Unit} is the number of parallel branches of one stator sub-module. The influence of the seam number N_{seam} and the seam width b_{seam} on the no-load motor performance is shown in Figure 5. It can be seen from Figure 5 that:

(1) The cogging torque amplitude was directly proportional to the seam number N_{seam} and the seam width b_{seam};

(2) The effective value of no-load phase back electromotive force, the amplitude of the self-inductance, and mutual inductance were inversely proportional to the seam number N_{seam} and the seam width b_{seam};

(3) The self-inductance L_{AA} and mutual inductance L_{AC} of A phase, which were closest to the seam, were the most affected by the seam number N_{seam} and the seam width b_{seam} because the maximum mutual inductance variation extent (0.35 mH), which is influenced by the seam number N_{seam} and the seam width b_{seam}, has little influence on motor performance; the mutual inductances that are influenced by the seam number N_{seam} and the seam width b_{seam} are ignored in the following chapters.

Table 1. Parameters of the 60-pole 72-slot motor.

Parameter	Values and Unit
Rated power of motor	220 kW
Rated voltage	660 V
Rated speed	30 r/min
Rated frequency	15 Hz
Number of rotor poles	60
Number of stator slots	72
Number of groups per pole per phase	12
Rotor structure	Out rotor with permanent magnet surface mounting
Outside diameter of rotor	1270 mm
Thickness of rotor yoke	50 mm
Core length	1000 mm
Thickness of PM	8 mm
Pole arc coefficient of PM	0.83
PM material	N42UH
Br at 20 °C	1.3T
Hc at 20 °C	971 kA/m
Relative magnet permeability	1.06
Air gap	2.3 mm
Outside diameter of stator	1149.4 mm
Slot height	80 mm
Stator yoke height	50 mm
Pitch	1
No-load back electromotive force	609.5 V

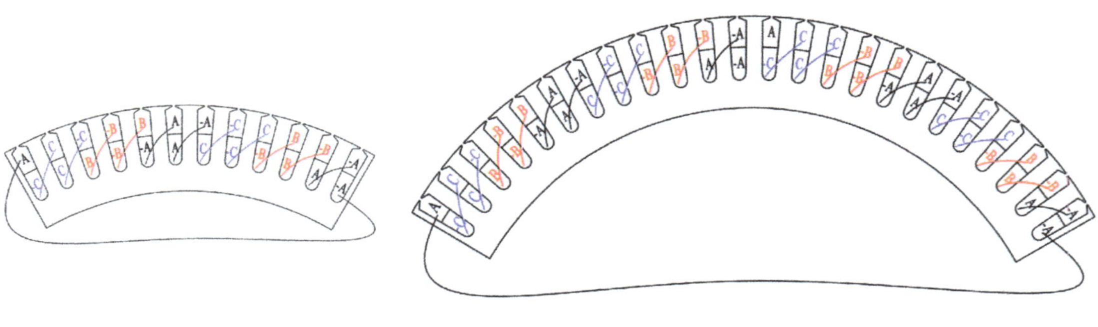

(a) $N_{seam} = 6$, a = 6, $a_{Unit} = 1$ **(b)** $N_{seam} = 3$, a = 3, $a_{Unit} = 1$

Figure 4. Single-stator unit winding end connection diagram with different N_{seam} values.

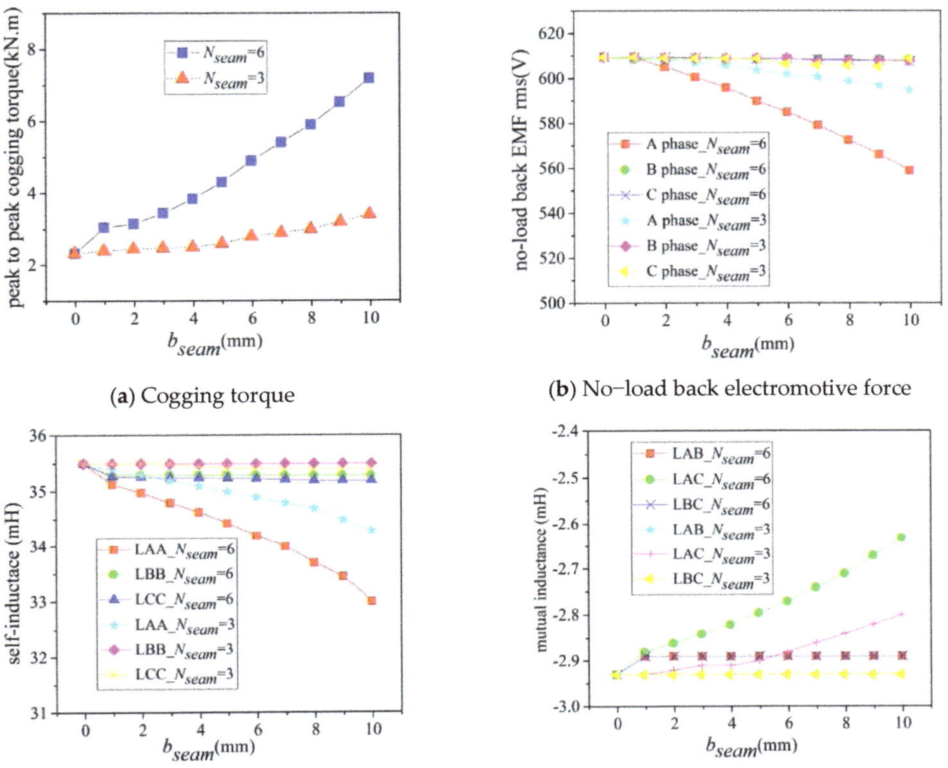

(**a**) Cogging torque

(**b**) No−load back electromotive force

(**c**) Self−inductance

(**d**) Mutual inductance

Figure 5. The relationships of no-load performance using different seam numbers and seam widths.

3.2. The Relationship between Motor No-Load Performance and the Number of Parallel Branches in One Unit with Same Seam

Winding end connection diagram with two parallel branches in one stator unit are shown in Figure 6, Where N_{seam} = 6, a = 12, a_{Unit} = 2. The no-load performance with a different number of parallel branches in one stator unit a_{Unit} is shown in Figure 7; the no-load cogging torque is not shown because the stator winding connection style had no influence on it. The winding end connection diagrams with a_{Unit} = 1 and a_{Unit} = 2 are shown in Figures 4a and 6, respectively.

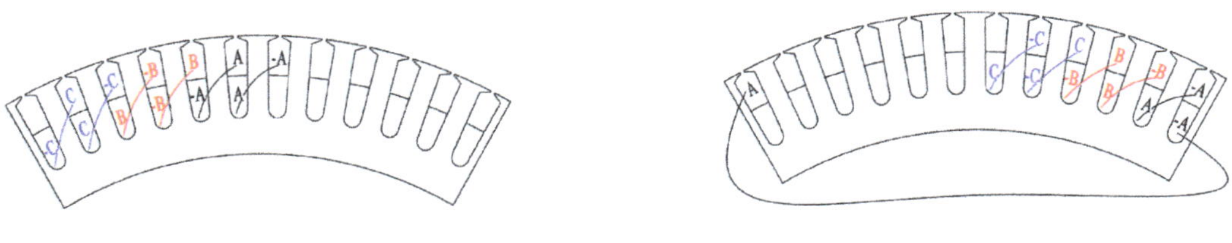

(**a**) Parallel branch 1 in one unit

(**b**) Parallel branch 2 in one unit

Figure 6. Winding end connection diagram with two parallel branches in one stator unit (N_{seam} = 6, a = 12, a_{Unit} = 2).

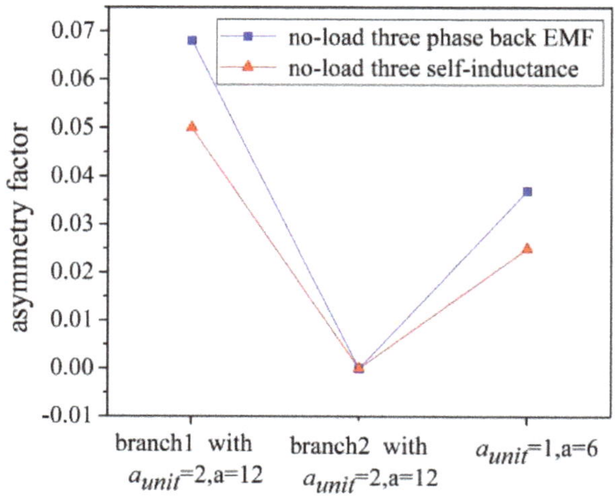

Figure 7. Asymmetry factor of no-load three-phase back electromotive force and self-induction using different motor parallel branches (b_{seam} = 5 mm).

Additionally, the asymmetry factor for the back electromotive force in this paper is $\max\{|E_A - E_{av}|; |E_B - E_{av}|; |E_C - E_{av}|\}$ and the asymmetry factor for self-inductance is $\max\{|L_{AA} - L_{av}|; |L_{BB} - L_{av}|; |L_{CC} - L_{av}|\}$, where E_A is the effective values of A phase winding no-load back electromotive force, E_B is the effective values of B phase winding no-load back electromotive force, E_C is the effective values of C phase winding no-load back electromotive force, $E_{av} = (E_A + E_B + E_C)/3$; L_{AA}, L_{BB}, L_{CC} are the average no-load self-inductance value of A phase winding, B phase winding, and C phase winding, respectively, and $L_{av} = (L_{AA} + L_{BB} + L_{CC})/3$.

It can be seen from Figure 7 that the asymmetry factor of the no-load three-phase back electromotive force's effective values and the inductance were inversely proportional to the number of parallel branches in one stator unit a_{Unit}.

3.3. The Relationship between Motor No-Load Performance and Slot–Pole Combination with Same Seam Parameters

The stator winding wiring diagrams with different pole slots are shown in Figure 8, where the winding wiring diagram of a 60-pole, 54-slot stator is divided into two kinds according to the position of the seam on the A phase winding. The no-load performance, which is shown in Figure 9, showed that:

(1) More stator slots with the same rotor poles can weaken the no-load motor performance impact of seams: with the same 54-slot stator sheet parameter, the no-load motor performance impact of a seam with the 36-pole scheme was less than the 60-pole scheme; with the same 60-pole rotor parameter, the no-load motor performance impact of a seam with a 72-slot scheme was less than that of a 54-slot scheme;

(2) When the number of stator teeth that continuously wind the same phase coil in one polar phase group was more than three, the seam on the side of the same phase coil tooth weakened the no-load motor performance impact of the seam. The no-load motor performance impact of the seam in Figure 8b was less than that in Figure 8c, where the seam is in the middle of the same phase coil tooth.

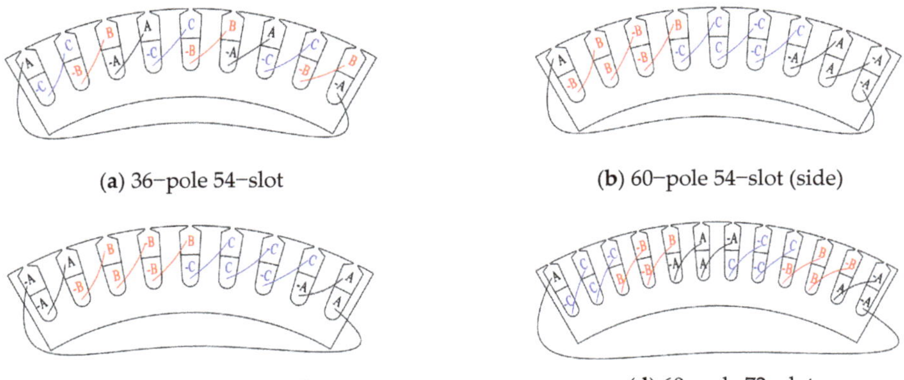

(a) 36–pole 54–slot (b) 60–pole 54–slot (side)

(c) 60–pole 54–slot (middle) (d) 60–pole 72–slot

Figure 8. Stator winding wiring diagram with different pole slots (N_{seam} = 6, a = 6, a_{Unit} = 1).

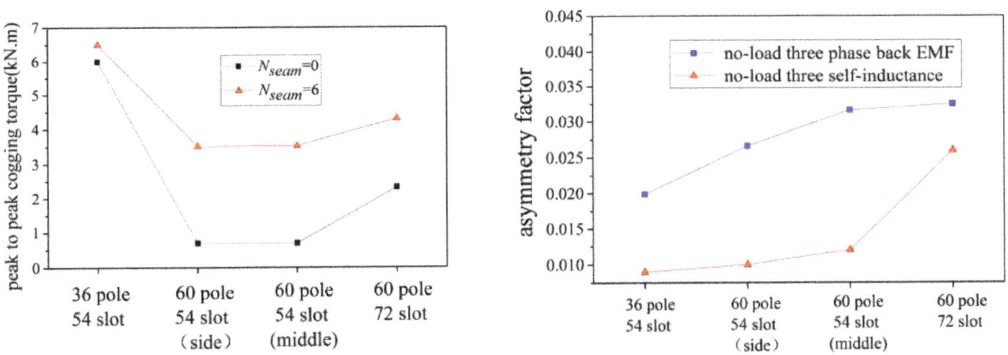

(a) Cogging torque (b) Back electromotive force and self–inductance

Figure 9. The relationship between no-load motor performance and slot–pole combination (b_{seam} = 5 mm).

In conclusion, the following electromagnetic parameters can weaken the no-load motor performance impact of seams:

(1) Reduce the number of the stator core seams and width of the stator core seams;
(2) Reduce the number of parallel branches in one stator sub-module;
(3) Adopt fewer poles and more slots;
(4) When the number of stator teeth that continuously wind the same phase coil in one polar phase group is more than three, place the seam on the same side as the phase coil tooth.

4. Stator Sheet Structures That Can Weaken the Impact of Seams

4.1. No-Load Motor Performance Impact of Seams on Different Parts of Tooth

Figure 10 shows a schematic drawing of a seam on different parts of tooth. The corresponding results, which are shown in Figure 11, showed that:

(1) The no-load cogging torque was chiefly affected by a seam on the top of the tooth (compare Figure 10a,c);
(2) The no-load back electromotive force and self-inductance were mainly affected by a seam on the top of the tooth (compare Figure 10a–c);
(3) The no-load motor performance impact of a seam on the yoke was negligible (compare Figure 10b,c).

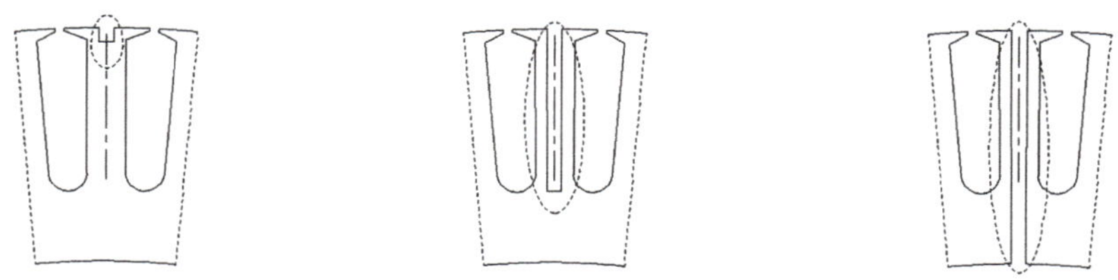

(**a**) Seam on the top of tooth (**b**) Seam on the tooth (**c**) Seam on the tooth and yoke

Figure 10. Schematic drawing of a seam on a stator core.

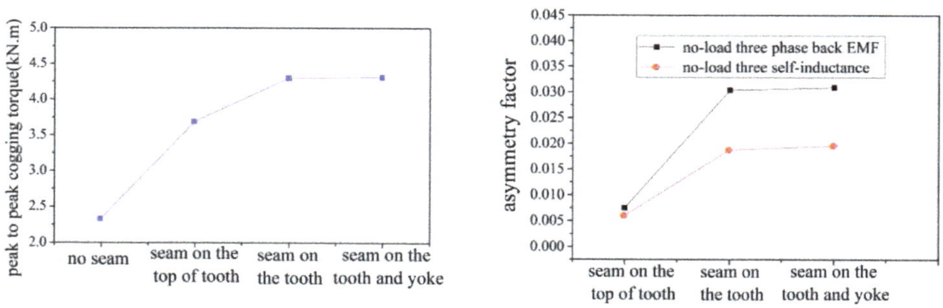

(**a**) Cogging torque (**b**) Back electromotive force and self–inductance

Figure 11. The corresponding no-load performance of seam location on stator core (N_{seam} = 6, b_{seam} = 5 mm, a = 6).

4.2. Proposed New Stator Sheet Structures

According to the conclusions from Section 3.1, the new stator sheet structures, which are shown in Figure 12, adopt the method of an opening auxiliary seam on the top of the tooth to balance the no-load cogging torque impact of a seam, and narrow the tooth or auxiliary seam on the whole tooth to balance the no-load back electromotive force and self-inductance. In Figure 12, the width of narrow tooth $b_{tz} = b_{t11} + b_{t12} = b_t - b_{seam}$.

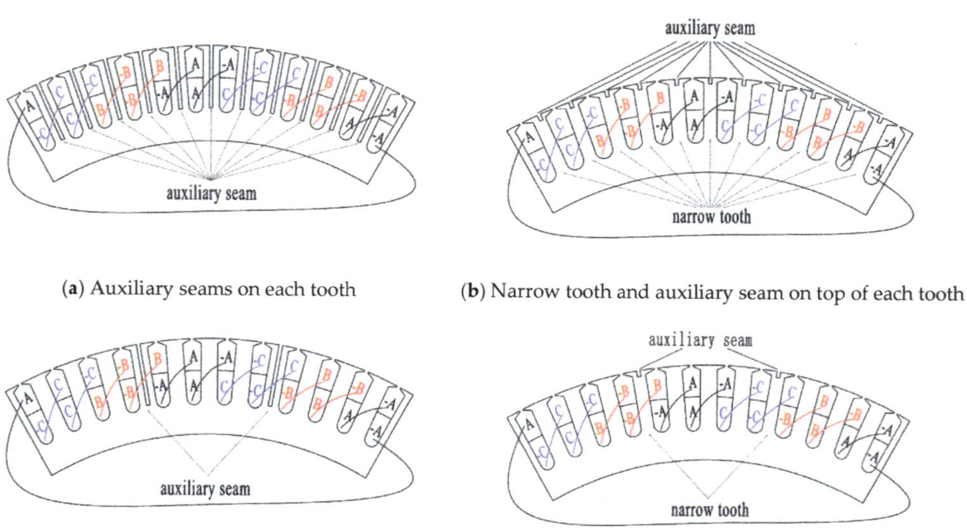

(**a**) Auxiliary seams on each tooth (**b**) Narrow tooth and auxiliary seam on top of each tooth

(**c**) 60–pole, 54–slot (middle) stator (**d**) 60–pole, 72–slot stator

Figure 12. Stator winding wiring diagrams with different pole slots (N_{seam} = 6, a = 6, a_{Unit} = 1).

Figure 13 shows that:

(1) Compared with the narrow tooth and auxiliary seam on the top of the tooth scheme (Figure 12b,d), an auxiliary seam (Figure 12a,c) was better at weakening the no-load cogging torque and self-inductance impact of the seam but worse at weakening the no-load asymmetry factor of the back electromotive force impact of the seam;

(2) Compared with one tooth of B and C phase with an auxiliary seam or narrow tooth (Figure 12c,d), each whole tooth with an auxiliary seam or narrow tooth (Figure 12a,b) weakened the no-load cogging torque and self-inductance impact of the seam better, but reduced the back electromotive force value more.

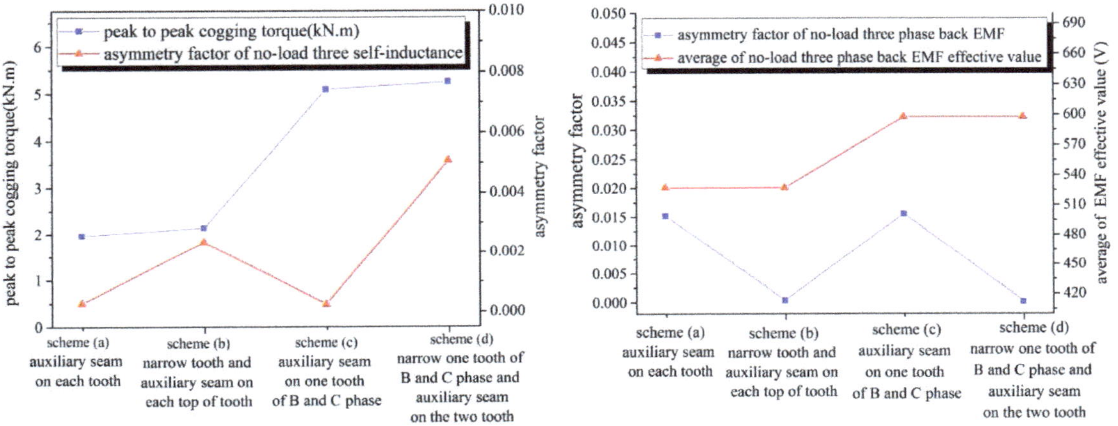

(**a**) Cogging torque and self–inductance (**b**) Back electromotive force

Figure 13. Comparison of no-load performance with auxiliary seam and narrow tooth (b_{seam} = 5 mm).

Learning from the strengths and weaknesses shown in Figure 12, Figure 14 was proposed on the basis of Figure 12b,d. An auxiliary seam 2 was added to balance the air-gap magnetic field to reduce the negative impact of the seam on the cogging torque, where the width of auxiliary seam 2 is equal to the width of auxiliary seam 1.

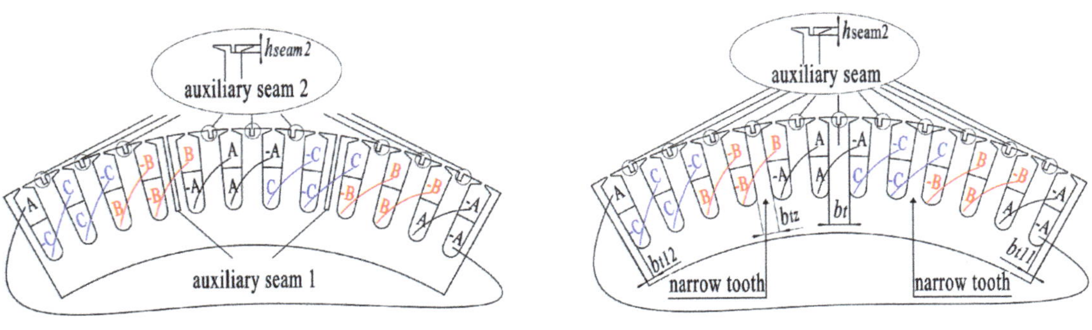

(**a**) (**b**)

Figure 14. Winding end connection diagram with double auxiliary seams scheme (N_{seam} = 6, a_{Unit} = 1). (**a**) Auxiliary seam 1 on one tooth of B and C phase and auxiliary seam 2 on other tooth, (**b**) Narrow tooth of B and C phase + auxiliary seam 1 on the narrow tooth + auxiliary seam 2 on the other tooth.

The results in Figures 15 and 16 show that:

(1) Auxiliary seam 2 can effectively weaken no-load cogging torque (compare with Figure 12b,d); meanwhile, auxiliary seam 2 weakened the no-load back electromotive force value less when compared with the results in Figure 12a,c;
(2) Increasing the height of auxiliary seam h_{seam2} can weaken the no-load cogging torque, but increase the asymmetry factor of no-load back electromotive force and decrease the value of no-load back electromotive force;
(3) When the seam width b_{seam} is increased (Figure 16), the cogging torque decreased to a smaller degree, and the asymmetry factor decreased, while no-load self-inductance and phase electromotive force remained nearly constant. Both schemes decreased the effective value of the back electromotive force to a greater degree, especially the Figure 14b scheme.

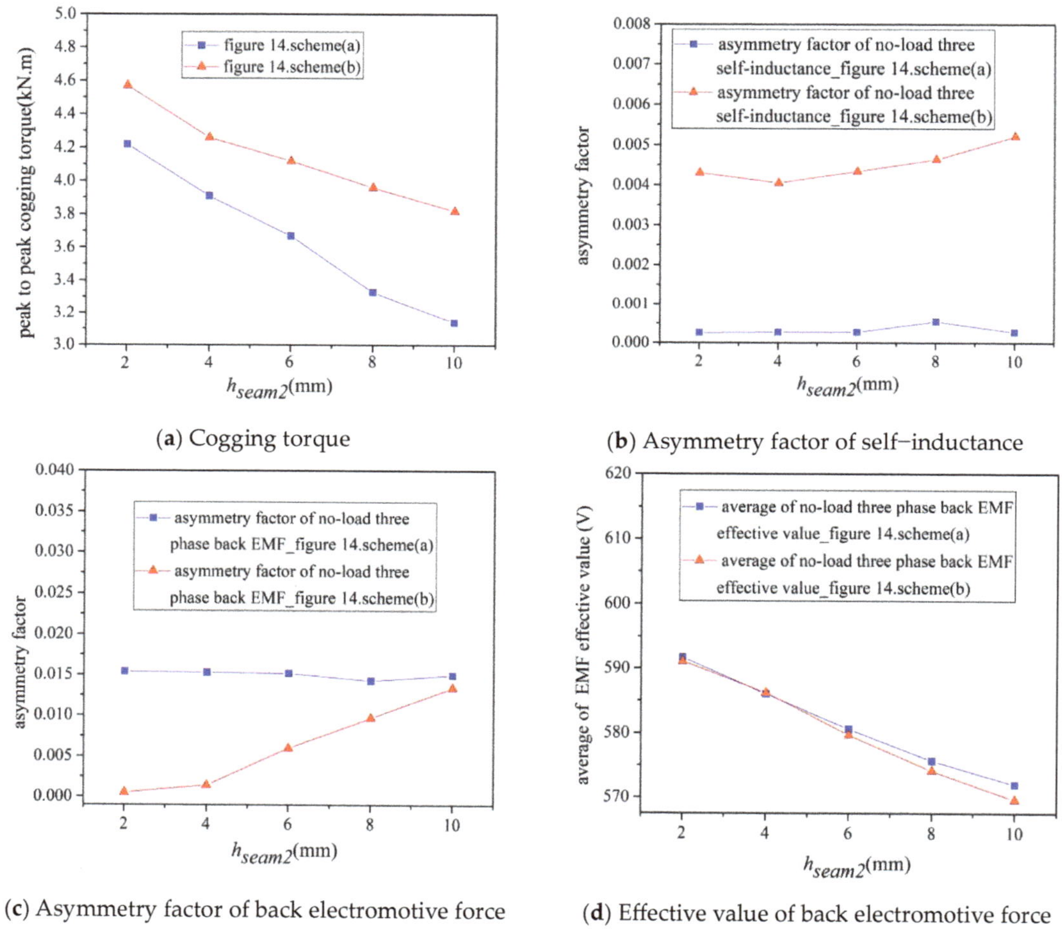

Figure 15. The corresponding no-load performance with double auxiliary seam scheme and different h_{seam2} (b_{seam} = 5 mm).

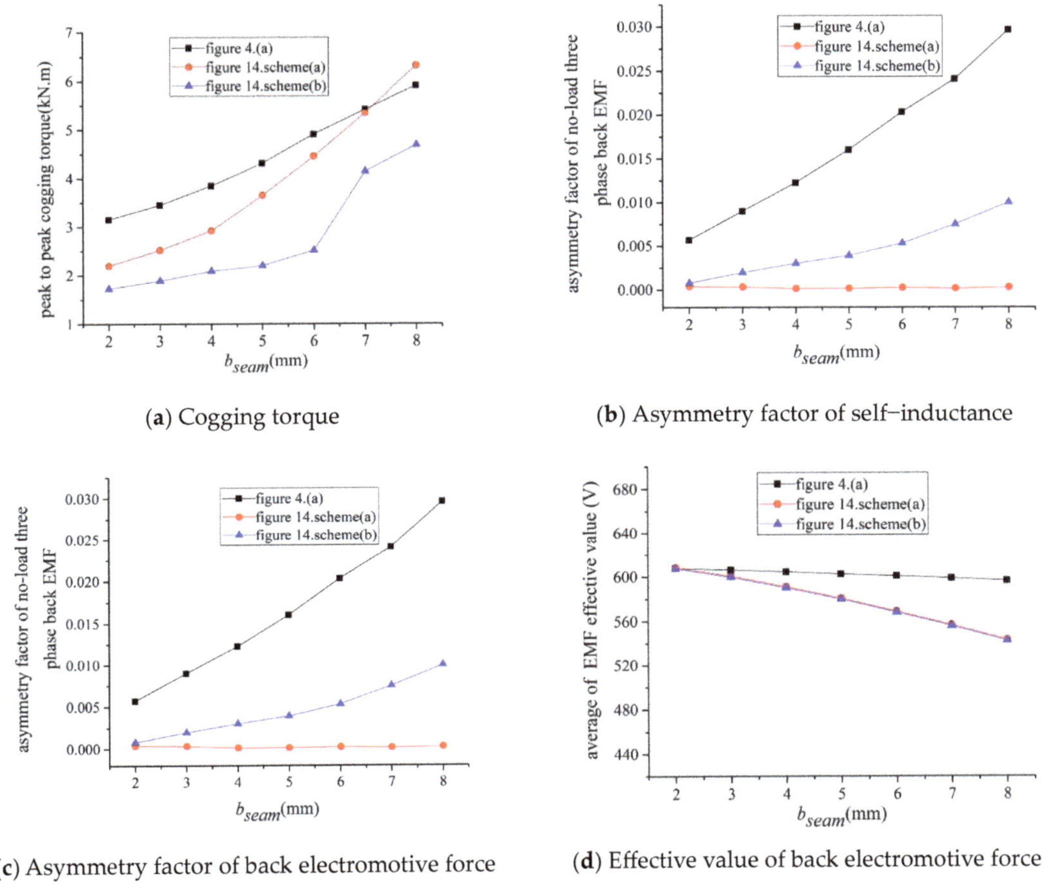

(**a**) Cogging torque

(**b**) Asymmetry factor of self–inductance

(**c**) Asymmetry factor of back electromotive force

(**d**) Effective value of back electromotive force

Figure 16. The corresponding no-load performance using double auxiliary seam scheme and different b_{seam} (h_{seam2} = 6 mm).

The disadvantage of the Figure 14a scheme is that auxiliary seam 1 produces difficulties on the core, especially if the seam width b_{seam} is too small. The disadvantage of the Figure 14b scheme is the effects on the asymmetry factor of no-load back electromotive force, especially if the seam width b_{seam} is too large. Therefore, when the seam width b_{seam} is larger, Figure 14a is the better scheme; when the seam width b_{seam} is smaller, Figure 14b is the better scheme.

5. Discussion

To model the impact of stator core seams on stator module-combined permanent magnet synchronous motors, a magnetic circuit model considering the stator core seam was deduced theoretically. The selection of electromagnetic parameters to weaken the impact of seams on the no-load performance is summarized as follows: (A) reduce the number and width of the seams; (B) reduce the number of parallel branches under the one stator sub-module; (C) adopt a scheme with fewer poles and more slots; and (D) when the number of stator teeth that continuously wind the same phase coil in one polar phase group is more than 3, place the seam on the same side as the phase coil tooth. The new stator core structures, which are shown in Figure 14, can significantly weaken the impact of seams on the no-load performance, while reducing the decreasing value of the no-load back electromotive force.

This study mainly focused on the seam on the stator core tooth; however, the influence of seams on the stator core yoke was not studied in this paper. The maximum modular number, arrangement of coils in the groove, and winding end wiring modes with different modular seam locations will be investigated in future studies.

Author Contributions: S.S. wrote the paper and implemented simulation; G.F. supervised all processes; B.Z. and Y.L. analyzed the data and checked paper format. All authors have read and agreed to the published version of the manuscript.

Funding: This research received no external funding.

Data Availability Statement: Data sharing not applicable.

Conflicts of Interest: The authors declare no conflict of interest.

References

1. Crider, J.M.; Sudhoff, S.D. An Inner Rotor Flux-Modulated Permanent Magnet Synchronous Machine for Low-Speed High-Torque Applications. *IEEE Trans. Energy Convers.* **2015**, *30*, 1247–1254. [CrossRef]
2. Valavi, M.; Le Besnerais, J.; Nysvee, A. An Investigation of Zeroth- Order Radial Magnetic Forces in Low-Speed Surface-Mounted Permanent Magnet Machines. *IEEE Trans. Magn.* **2016**, *52*, 1–6. [CrossRef]
3. Fan, Y.; Zhu, W.; Zhang, X.; Cheng, M.; Chau, K.T. Research on a Single Phase-Loss Fault-Tolerant Control Strategy for a New Flux-Modulated Permanent-Magnet Compact In-Wheel Motor. *IEEE Trans. Energy Convers.* **2016**, *31*, 658–666. [CrossRef]
4. Li, G.-J.; Zhu, Z.-Q.; Foster, M.P.; Stone, D.A.; Zhan, H.-L. Modular Permanent Magnet Machines with Alternate Teeth Having Tooth Tips. *IEEE Trans. Ind. Electron.* **2015**, *62*, 6120–6130. [CrossRef]
5. Wang, J.; Atallah, K.; Zhu, Z.Q.; Howe, D. Modular Three-Phase Permanent-Magnet Brushless Machines for In-Wheel Applications. *IEEE Trans. Veh. Technol.* **2008**, *57*, 2714–2720. [CrossRef]
6. Zhu, Z.Q.; Thomas, A.S.; Chen, J.T.; Jewell, G.W. Cogging torque in flux-switching permanent magnet machines. *IEEE Trans. Magn.* **2009**, *45*, 4708–4711. [CrossRef]
7. Zhu, Z.Q.; Chen, J.T.; Wu, L.J.; Howe, D. Influence of stator asymmetry on cogging torque of permanent magnet brushless machines. *IEEE Trans. Magn.* **2008**, *44*, 3851–3854. [CrossRef]
8. Li, G.J.; Zhu, Z.Q. Analytical Modeling of Modular and Unequal Tooth Width Surface-Mounted Permanent Magnet Machines. *IEEE Trans. Magn.* **2015**, *51*, 8107709. [CrossRef]
9. Gan, B.; Zhang, B.; Li, Q.; Guihong, F.; Li, G. Research on Operation of Low-Speed and High-Torque Module Combined Stator Permanent Magnetic Fault-Tolerant Motor With Unequal Span Winding. *IEEE Access* **2020**, *8*, 166824–166838. [CrossRef]
10. Xu, Y.; Zhang, B.; Feng, G. Analysis of Unwinding Stator Module Combined Permanent Magnet Synchronous Machine. *IEEE Access* **2020**, *8*, 191901–191909. [CrossRef]
11. Zhang, B.; Gan, B.; Li, Q. Analysis of a Fault-Tolerant Module-Combined Stator Permanent Magnet Synchronous Machine. *IEEE Access* **2020**, *8*, 70438–70452. [CrossRef]
12. Liu, Y.; Zhang, B.; Zong, M.; Feng, G.; Gan, B. Magnetic Field Prediction of Module-Combined Stator Permanent Magnet Synchronous Motor Based on a Nonlinear Hybrid Analytical Model. *IEEE Access* **2021**, *9*, 122486–122494. [CrossRef]
13. Li, G.J.; Ren, B.; Zhu, Z.Q.; Li, Y.X.; Ma, J. Cogging Torque Mitigation of Modular Permanent Magnet Machines. *IEEE Trans. Magn.* **2016**, *52*, 8100210. [CrossRef]
14. Li, G.J.; Ren, B.; Zhu, Z.Q. Cogging torque and torque ripple reduction of modular permanent magnet machines. In Proceedings of the 2016 XXII International Conference on Electrical Machines (ICEM), Lausanne, Switzerland, 4–7 September 2016; pp. 193–199. [CrossRef]
15. Li, G.J.; Zhu, Z.Q.; Chu, W.Q.; Foster, M.P.; Stone, D.A. Influence of Flux Gaps on Electromagnetic Performance of Novel Modular PM Machines. *IEEE Trans. Energy Convers.* **2014**, *29*, 716–726. [CrossRef]
16. Laldin, O.; Sudhoff, S.D.; Pekarek, S. Modified Carter's Coefficient. *IEEE Trans. Energy Convers.* **2015**, *30*, 1133–1134. [CrossRef]
17. Yin, H.; Zhang, H.; Hua, W.; Su, P. Improved Open-Circuit Airgap Field Model for FSCW-STPM Machines Considering PM-MMF Fluctuation. *IEEE Trans. Ind. Electron.* **2022**, *69*, 5547–5556. [CrossRef]
18. Krause, P.; Wasynczuk, O.; Sudhoff, S.D.; Pekarek, S. Appendix B: Carter's Coefficient. In *Analysis of Electric Machinery and Drive Systems*; Wiley-IEEE Press: New York, NY, USA, 2013; pp. 626–628. [CrossRef]

energies

Article

Analysis of Torque Ripple in V-Shape Interior Permanent Magnet Machine Based on General Airgap Field Modulation Theory

Jiawei Zhou, Ming Cheng *, Wenfei Yu and Wei Hua

School of Electrical Engineering, Southeast University, Nanjing 210000, China; jwzhou@seu.edu.cn (J.Z.); wenfeiyu@seu.edu.cn (W.Y.); huawei1978@seu.edu.cn (W.H.)
* Correspondence: mcheng@seu.edu.cn

Abstract: This paper proposes a novel torque ripple model of V-Shape interior permanent magnet (IPM) synchronous machine based on the general field modulation theory (GAFMT). Firstly, the magnetic field modulation behavior of a 12-stator-slot/8-rotor-pole (12/8) IPM machine is discussed, where the expressions of permanent-magnet-excited (PM-excited) and armature fields are derived. Secondly, the total torque ripple is divided into three components, namely, the PM-excited cogging torque, the armature cogging torque, and the modulation torque ripple. Additionally, the relationship between the field harmonics and the torque ripple is discussed based on the GAFMT. Then, a finite element analysis (FEA)-assisted dual fast Fourier transformation (FFT) method is proposed to obtain the accurate value of the field amplitude for torque ripple calculation. The field analysis results based on the GAFMT and the proposed torque ripple are then validated by comparing their results with those obtained from the FEA simulations, and a good agreement is observed, demonstrating the effectiveness of the field analysis and the proposed torque ripple model. Finally, experiments on a prototype 12/8 IPM machine demonstrate the validity of the torque ripple model, providing an effective method for torque ripple suppression of IPM machines.

Keywords: airgap field modulation; interior permanent magnet machine; torque ripple; analytical method

Citation: Zhou, J.; Cheng, M.; Yu, W.; Hua, W. Analysis of Torque Ripple in V-Shape Interior Permanent Magnet Machine Based on General Airgap Field Modulation Theory. *Energies* **2023**, *16*, 4586. https://doi.org/10.3390/en16124586

Academic Editors: Frede Blaabjerg and Angelo Maiorino

Received: 8 May 2023
Revised: 31 May 2023
Accepted: 6 June 2023
Published: 8 June 2023

1. Introduction

V-Shape interior permanent magnet (IPM) synchronous machine is a typical topology with a wide range of applications in servo [1], electric vehicle [2], and industrial robotics [3] due to its high power density, high efficiency, and wide speed range. The torque ripple of IPM machines is a major concern in these high-quality applications, which will produce mechanical vibration and reduce the motion performance [4].

The mechanism of torque ripple generation of IPM machines is complex and still unclear because it is related to the time and space harmonics of magnetic fields due to permanent magnet (PM) and armature winding [5]. Therefore, to study the mechanism of torque ripple, the first step is to analyze the magnetic field characteristics in the airgap. However, conventional electrical machine theories, such as the cross-field theory for AC machines [6], the magnetic equivalent circuit theory for regular machines [7], and the rotating magnetic field theory [8], are usually applied to model the fundamental magnetic field while neglecting the harmonic field.

Recently, a general airgap field modulation theory (GAFMT) was proposed [9], which models all the electrical machines in a generalized way by three key components, namely the excitation source, the modulator, and the filter (armature winding), and thus, the characteristics of various magnetic field harmonics can be investigated. The GAFMT can not only be used to analyze the performance of "non-conventional" machines [10], e.g., flux switching PM (FSPM) machine [11], vernier machine (VM) [12], and flux-reversal PM machine (FRPM) [13], etc., but also conventional machines [14,15], providing new ideas for machine design and analysis.

While the GAFMT has been proved to be effective in the analysis of spatial distribution of field harmonics [16], it still has flaws in the amplitude analysis of field harmonics, since the permeability function is usually simplified in the analysis process. The finite element analysis (FEA) method can deal with the influence of complex structures and slotting effects on the magnetic fields and therefore significantly improve the field amplitude analysis accuracy [2,17,18]. Hence, to further improve the ability to analyze the amplitude of field harmonics for torque ripple calculation, this paper proposes an FEA-based dual fast Fourier transformation (FFT) method to extract the amplitudes from the magnetic fields.

Further, based on the combination of GAFMT and FEA-based analysis method, an accurate torque ripple model for IPM machines is proposed, which allows an intuitive description of the torque ripple generation mechanism and quantifies the contribution of each magnetic field harmonic to the total torque ripple. In addition, since the maximum torque per ampere (MTPA) method is usually applied in the IPM machines, the effect of the armature current phase on torque ripple is also investigated based on the proposed torque ripple model.

The organization of the rest of this paper is as follows. The field modulation behavior of IPM machines is analyzed in Section 2, and the torque ripple model and the relationship between the field harmonics and the torque ripple are discussed in Section 3. Then, an FEA-assisted dual FFT method is proposed in Section 4 to derive the detailed torque ripple model. Then, simulations on a 12/8 IPM machine are conducted through FEA together with a comparison with the results from the proposed model in Section 5. The experimental verifications are presented in Section 6, followed by the conclusion drawn in Section 7.

2. Field Modulation Behavior of V-Shape IPM Machine

From the GAFMT perspective, the V-Shape IPM machine can be considered as a class of salient pole reluctance motor, both in terms of its excitation and armature fields modulated by the salient stator/rotor pole. However, the field modulation behavior is not exactly the same for both fields. In the following part, the modulation behavior of the V-Shape IPM is introduced in detail with the aid of the GAFMT.

2.1. Rotor Modulation of PM-Excited Field

To illustrate the rotor modulation behavior for the PM-excited field, the stator is considered as a non-slot structure temporarily, and the rotor modulator structure seen from the PM-excited field is shown in Figure 1, where ε_{rt1} is the pole–arc ratio, N_{rt} is the rotor pole number, T_{rt} (rad) is the radian of unit rotor modulator ($2\pi/N_{rt}$). As PMs are installed in the V-Shape slots, most of the PM fluxes flow from the silicon steel within the clamping angle of PMs to the stator silicon steel through airgap, and the remaining PM fluxes flow to adjacent PMs through the flux leakage path provided by the magnetic bridge. Hence, there are few PM fluxes in the airgap area corresponding to the bridge. The distribution function of the PM flux after the modulation of the rotor pole can be described as a segmental function in Equation (1), as shown in Figure 2.

$$M_{rt}[f_m(\phi)] = \begin{cases} 0, & -\frac{2\pi}{N_{rt}} < \phi < -\frac{\pi}{N_{rt}} - \varepsilon_{rt1}T_{rt} \\ -1, & -\frac{\pi}{N_{rt}} - \varepsilon_{rt1}T_{rt} < \phi < -\frac{\pi}{N_{rt}} + \varepsilon_{rt1}T_{rt} \\ 0, & -\frac{\pi}{N_{rt}} + \varepsilon_{rt1}T_{rt} < \phi < \frac{\pi}{N_{rt}} - \varepsilon_{rt1}T_{rt} \\ 1, & \frac{\pi}{N_{rt}} - \varepsilon_{rt1}T_{rt} < \phi < \frac{\pi}{N_{rt}} + \varepsilon_{rt1}T_{rt} \\ 0, & \frac{\pi}{N_{rt}} + \varepsilon_{rt1}T_{rt} < \phi < \frac{2\pi}{N_{rt}} \end{cases} \quad (1)$$

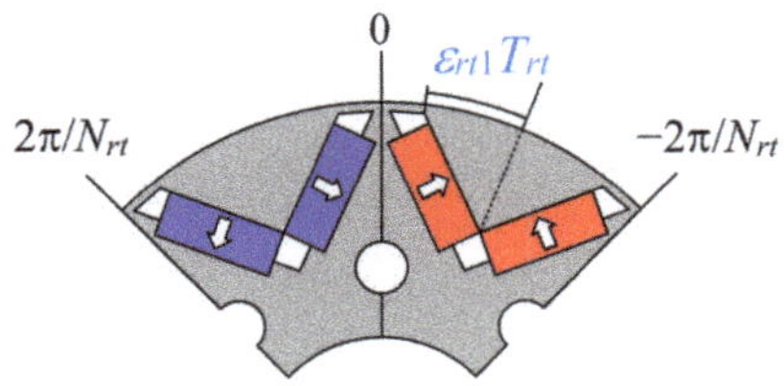

Figure 1. Rotor modulator structure for PM-excited field.

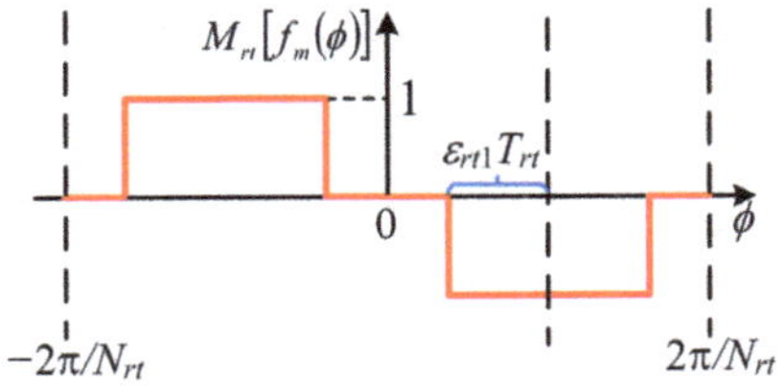

Figure 2. Distribution of PM-excited field.

Considering the rotation of the rotor modulator, Equation (1) can be transformed into the Fourier series form as

$$M_{rt}[f_m(\phi,t)] = \sum_{v}^{\infty} C_{mr}(v)\sin\left(v\frac{N_{rt}}{2}(\phi-\omega t)\right) \tag{2}$$

where ϕ is the mechanical angle along the circumference in the modulator-fixed reference frame. $M_{rt}[f_m(\phi,t)]$ represents the circumference distribution of the rotor-modulated PM-excited field. v is a positive integer, which represents the modulation order. ω (rad/s) is the mechanical rotation speed. t (s) is time, and C_{mr} is the rotor modulation coefficient, which can be described as

$$C_{mr}(v) = -\frac{4}{\pi v}\sin(2\pi v\varepsilon_{rt})\sin\left(\frac{\pi v}{2}\right) \tag{3}$$

2.2. Stator Modulation of PM-Excited Field

The stator modulator structure for the PM-excited field is shown in Figure 3a, and the operator function shown in Figure 3b can be described as

$$M_{st}[\cdot] = 2\varepsilon_{st} + \sum_{l}^{\infty} \frac{2}{\pi l}\sin(2\pi l\varepsilon_{st})\cos(lN_{st}\phi) \tag{4}$$

where l is a positive integer, which represents the modulation order. N_{st} is the number of the stator tooth. T_{st} is the radian of the unit stator modulator ($2\pi/N_{st}$). ε_{st} is half of the pole–arc ratio of stator tooth, and $M_{st}[\cdot]$ represents the stator modulator operator.

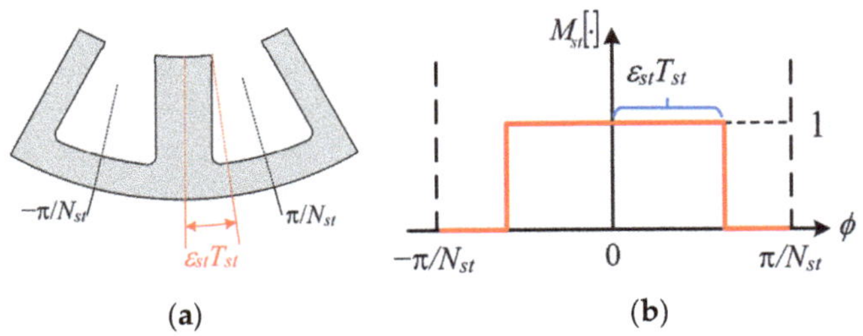

Figure 3. Stator modulator for PM-excited field: (**a**) Structure; (**b**) Operator function.

2.3. Stator Modulation of Armature Field

The winding function shown in Figure 4 can be described as

$$\begin{cases} W_a(\phi) = \sum_v^\infty \frac{2}{\pi v} \sin\left(\frac{\pi v}{3}\right) \cos(vp\phi) \\ W_b(\phi) = \sum_v^\infty \frac{2}{\pi v} \sin\left(\frac{\pi v}{3}\right) \cos\left(vp\left(\phi - \frac{2\pi}{3}\right)\right) \\ W_c(\phi) = \sum_v^\infty \frac{2}{\pi v} \sin\left(\frac{\pi v}{3}\right) \cos\left(vp\left(\phi + \frac{2\pi}{3}\right)\right) \end{cases} \quad (5)$$

where p is the rotor pole pairs. W_a, W_b, and W_c represent the winding function of phase A, B, and C respectively.

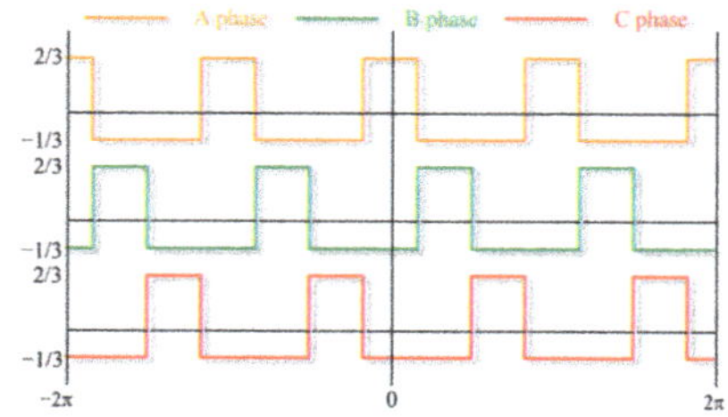

Figure 4. Distribution of winding function.

Considering the armature current as

$$\begin{cases} I_a = \sum_\delta^\infty I_m^\delta \sin(\delta p\omega t) \\ I_b = \sum_\delta^\infty I_m^\delta \sin\left(\delta\left(p\omega t - \frac{2\pi}{3}\right)\right) \\ I_c = \sum_\delta^\infty I_m^\delta \sin\left(\delta\left(p\omega t + \frac{2\pi}{3}\right)\right) \end{cases} \quad (6)$$

where δ is a positive integer, which represents the order of current harmonics. I_m^δ is the current amplitude of δth harmonic. I_a, I_b, and I_c represent the armature current of phase A, B, and C, respectively.

Multiplying Equation (5) by Equation (6), the stator-modulated armature excited field can be described as

$$M_{st}[f_a(\phi, t)] = \sum_\delta^\infty \sum_v^\infty C_{as}(v, \delta) \sin(vp\phi \pm \delta p\omega t) \quad (7)$$

where C_{as} is

$$C_{as}(v, \delta) = \left(\mp\frac{3}{2}\right)\left(\frac{2}{\pi v}\right) \sin\left(\frac{\pi v}{3}\right) I_m^\delta$$

2.4. Rotor Modulation of Armature Field

Figure 5 shows the unit rotor modulator structure and the modulator operator function. The modulator can be divided into three regions along the airgap, as shown in Figure 5a. The armature field fluxes tend to pass through the airgap area corresponding to regions 1 and 3, since the magnetic bridge is thin and usually saturated by PM fluxes. Thus, the rotor modulator function can be described in Figure 5b. The positive half-period expression of the modulation operator in Figure 5b is

$$M_{rt}[\cdot] = \begin{cases} 1, & -\frac{T_{rt}}{2} < \phi < -\varepsilon_{rt2} T_{rt} \\ \tau_r, & -\varepsilon_{rt2} T_{rt} < \phi < -\varepsilon_{rt1} T_{rt} \\ 1, & -\varepsilon_{rt1} T_{rt} < \phi < \varepsilon_{rt1} T_{rt} \\ \tau_r, & \varepsilon_{rt1} T_{rt} < \phi < \varepsilon_{rt2} T_{rt} \\ 1, & \varepsilon_{rt2} T_{rt} < \phi < \frac{T_{rt}}{2} \end{cases} \quad (8)$$

where T_{rt} denotes the radian of the unit rotor modulator. ε_{rt1} and ε_{rt2} denote the radian from the inner and outer edge of the magnetic bridge to the center of the rotor modulator, respectively, and $M_{rt}[\cdot]$ denotes the rotor modulation operator. The effect of the magnetic bridge on rotor modulator permeability is described by introducing a coefficient τ_r related to the thickness and saturation degree of the magnetic bridge. In this paper, the coefficient τ_r is obtained by FEA simulations. Specifically, the value of τ_r is approximated by comparing the armature field distribution with and without the rotor modulator and analyzing the differences in the magnetic density values at the corresponding positions in region 2.

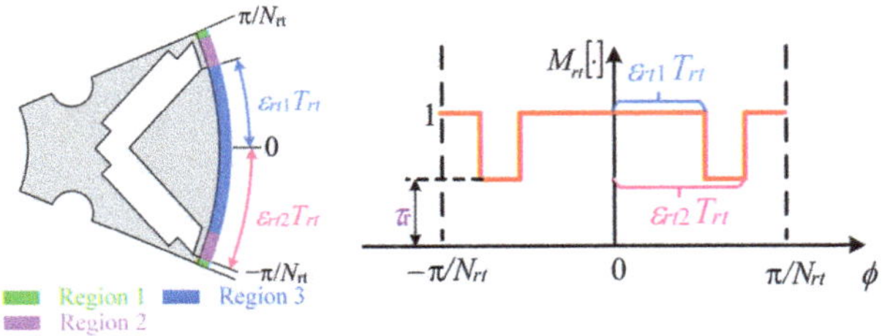

(a) (b)

Figure 5. Rotor modulator for Armature field: (**a**) Structure; (**b**) Operator function.

Considering the rotation of the rotor modulator, Equation (8) is transformed into the Fourier series form as

$$M_{rt}[\cdot] = \varepsilon_r + \sum_l^{\infty} C_{ar}(l)\sin(2lp(\phi - \omega t)) \tag{9}$$

where l represents the rotor modulation order. C_{ar} represents the harmonic amplitude, and ε_r represents the DC component of the Fourier series of the rotor modulator; the detailed expressions of them are

$$\begin{aligned} &\varepsilon_r = 1 + 2(1-\tau_r)(\varepsilon_{rt1} - \varepsilon_{rt2}) \\ &C_{ar}(l) = \tfrac{4}{\pi l}(1-\tau_r)\cos(l\pi(\varepsilon_{rt1} + \varepsilon_{rt2}))\sin(l\pi(\varepsilon_{rt1} - \varepsilon_{rt2})) \end{aligned}$$

2.5. Modulated PM-Excited and Armature Fields

The modulated PM-excited field can be expressed by multiplying Equations (2) and (4) as

$$F_m(\phi,t) = \begin{cases} \sum_v^{\infty} C_{mp}(v)\sin(vp\phi - vp\omega t) \\ +\sum_l^{\infty}\sum_v^{\infty} C_{msum}(v,l)\sin((vp + lN_{st})\phi - vp\omega t) \\ +\sum_l^{\infty}\sum_v^{\infty} C_{mdif}(v,l)\sin((vp - lN_{st})\phi - vp\omega t) \end{cases} \tag{10}$$

where C_{mp}, C_{msum}, C_{mdif} represent the coefficient of primitive, summation, and differential modulation processes, respectively, where

$$\begin{cases} C_{mp}(v) = \varepsilon_{st}\frac{8}{\pi v}\sin(2\pi v\varepsilon_{st})\sin(\frac{\pi v}{2}) \\ C_{msum}(v,l) = C_{mdif}(v,l) = \frac{2}{\pi v}\frac{2}{\pi l}\sin(2\pi l\varepsilon_{st})\sin(2\pi v\varepsilon_{st})\sin(\frac{\pi v}{2}) \end{cases}$$

Similarly, by multiplying Equations (7) and (9), the modulated armature field can be expressed as

$$F_a(\phi,t) = \begin{cases} \sum_\delta^{\infty}\sum_v^{\infty} C_{ap}(v,\delta)\sin(vp\phi \pm \delta p\omega t) \\ +\sum_\delta^{\infty}\sum_l^{\infty}\sum_v^{\infty} C_{asum}(v,\delta,l)\sin((vp + 2lp)\phi + (\pm\delta - 2l)p\omega t) \\ +\sum_\delta^{\infty}\sum_l^{\infty}\sum_v^{\infty} C_{adif}(v,\delta,l)\sin((vp - 2lp)\phi + (\pm\delta - 2l)p\omega t) \end{cases} \tag{11}$$

where

$$\begin{cases} C_{ap}(v,\delta) = \varepsilon_r C_{as}(v,\delta) \\ C_{asum}(v,\delta,l) = C_{adif}(v,\delta,l) = \frac{C_{as}(v,\delta)C_{ar}(l)}{2} \end{cases}$$

3. Torque Ripple Analysis

3.1. Three Torque Ripple Components

The virtual work method is an effective way to calculate electromagnetic (EM) torque and reveals the contribution of each field harmonic. Torque ripple can be considered to be caused by the change in magnetic co-energy in the airgap with the rotor position, so that torque ripple can be obtained from the derivation of magnetic co-energy with respect to the rotor position, as expressed by

$$T = \frac{\partial \int_V w_m dV}{\partial \omega t} = \frac{1}{2\mu_0}\frac{\partial \int_V B^2 dV}{\partial \omega t} \tag{12}$$

where B denotes the flux density, and μ_0 denotes the vacuum permeability.

The total flux density is a combination of PM-excited and armature flux densities, as follows:

$$B^2 = (B_m + B_a)^2 = B_m^2 + B_a^2 + 2B_m B_a \tag{13}$$

Thus, Equation (12) can be further expressed as

$$T = \frac{1}{2\mu_0}\frac{\partial \int_V B_m^2 dV}{\partial \omega t} + \frac{1}{2\mu_0}\frac{\partial \int_V B_a^2 dV}{\partial \omega t} + \frac{1}{2\mu_0}\frac{\partial \int_V 2B_m B_a dV}{\partial \omega t} \tag{14}$$

All of the three terms in Equation (14) might generate ripple torque, and thus, they are respectively referred to as PM-excited cogging torque, armature cogging torque, and modulation torque ripple based on their sources. Then, Equation (14) can be expressed as

$$T_{ripple} = T_{mcog} + T_{acog} + T_{modu} \tag{15}$$

where T_{ripple}, T_{mcog}, T_{acog}, T_{modu} denote the total torque ripple, the PM-excited cogging torque, the armature-excited cogging torque, and the modulation torque ripple, respectively.

The third term in Equation (15) is named as the modulation torque ripple because the source of it is related to the modulation behavior, which will be further discussed in the following section.

3.2. Relationship between the Field Harmonics and Torque Ripple

3.2.1. PM-Excited Cogging Torque and Armature Cogging Torque

Take the PM-excited field as an example, according to Equation (10), when the pole pairs of primitive-modulated field harmonic are the same as that of summation-modulated field harmonic as

$$v_p p = v_{sum} p + l_{sum} N_{st} = \alpha \tag{16}$$

where v_p, v_{sum} represent the rotor modulation order of primitive modulation and summation modulation, respectively. l_{sum} represents the stator modulation order of summation modulation. α is an odd positive integer.

The synthetic field harmonic can be expressed as

$$B_s = A_p \sin(\alpha\phi - \beta_1 \omega t) + A_{sum} \sin(\alpha\phi - \beta_2 \omega t) \tag{17}$$

where A_p, A_{sum} represent the amplitude of primitive-modulated and summation-modulated field harmonic, respectively. B_s represents the synthetic field harmonic with α pole pairs. β_1, β_2 represent the temporal order of primitive-modulated and summation-modulated field harmonic, respectively.

Equation (17) can be further expressed as

$$B_s = A_s \sin\left(\alpha\phi + \frac{(\beta_1 - \beta_2)\omega t}{2} + \varphi\right) \tag{18}$$

where A_s represents the amplitude of synthetic field harmonic, φ represents the phase shift according to the synthesis, and expressed as

$$A_s = \sqrt{A_p^2 + A_{sum}^2 + 2A_p A_{sum} \cos((\beta_1 + \beta_2)\omega t)} \tag{19}$$

$$\tan\varphi = \frac{A_p - A_{sum}}{A_p + A_{sum}} \tan\left(\frac{(\beta_1 + \beta_2)\omega t}{2}\right) \tag{20}$$

It can be concluded that the amplitude of the synthetic field harmonic is related to the source field harmonics and ripple when the rotor is rotated.

Substituting Equation (18) into the first term in Equation (14), the PM-excited cogging torque can be expressed as

$$T_{mcog} = \frac{\pi(R_s^2 - R_r^2)l_{stk}}{4\mu_0} \frac{\partial A_s}{\partial \omega t} \tag{21}$$

where R_s represents the outer radius of the stator. R_r represents the inner radius of the rotor.

Clearly, the PM-excited cogging torque is directly related to the amplitude of synthetic field harmonic, as is the armature cogging torque. Hence, the mechanisms of PM-excited cogging torque and armature cogging torque can be revealed as follows: due to the field modulation effect, the pole pairs of the different modulated field harmonics might be the same, but the rotation speed of them is not; this phenomenon results in the amplitude ripple of the synthetic field harmonic and further generates the torque ripple.

3.2.2. Modulation Torque Ripple

Similarly, the modulation torque ripple is related to the modulation effect; the characteristics of the PM-excited field and the armature field are shown in Table 1. When the rotation speeds of PM-excited field and armature field with the same pole pairs are the same, the constant output torque is produced; otherwise, the ripple torque is produced. Since this kind of ripple torque is determined by the modulation difference of the PM-excited field and armature field, it is referred to as the modulation torque ripple.

Table 1. Characteristics of the PM-excited field and the armature field.

Item	PM-Excited Field		Armature Field	
	Pole Pairs	Rotation Speed	Pole Pairs	Rotation Speed
Primitive modulation	vp	$vp\omega$	vp	$\pm\delta p\omega$
Summation modulation	$vp + lN_{st}$	$vp\omega$	$vp + 2lp$	$(\pm\delta - 2l)p\omega$
Differential modulation	$vp - lN_{st}$	$vp\omega$	$vp - 2lp$	$(\pm\delta + 2l)p\omega$

4. FEA-Assisted Dual FFT Analysis Method and Torque Ripple Model

4.1. FEA-Assisted Dual FFT Analysis Method

Based on the above analysis, the key point of calculating the torque ripple is obtaining the amplitude of field harmonics; hence, an FEA-assisted dual FFT analysis method is proposed in this section. This method is based on the GAFMT and assisted by the FEA. In this method, the GAFMT is mainly used in the first spatial FFT decomposition process to quickly select effective field harmonics, while the FEA allows a more precise quantification of the field harmonic amplitude ripple. Compared with the fully GAFMT-based method, the proposed method can save the convolution process while obtaining more accurate ripple values.

The key step of the proposed method is the FEA-based analysis process. The detailed flow chart is shown in Figure 6. Firstly, the effective field harmonics are selected by using the GAFMT in order to exclude invalid field harmonics and save time in the following steps. Then, the FEA simulation is used to obtain the flux density distribution in an electrical period. Furthermore, a first-round FFT analysis is carried out to calculate the amplitude of spatial field harmonics. On this basis, the harmonic coefficient of field amplitudes can be extracted from the second-round FFT analysis. Because the amplitude of field harmonics varies with the rotor position, which is time dependent, it is a function of time, and this process is essentially a temporal FFT decomposition process.

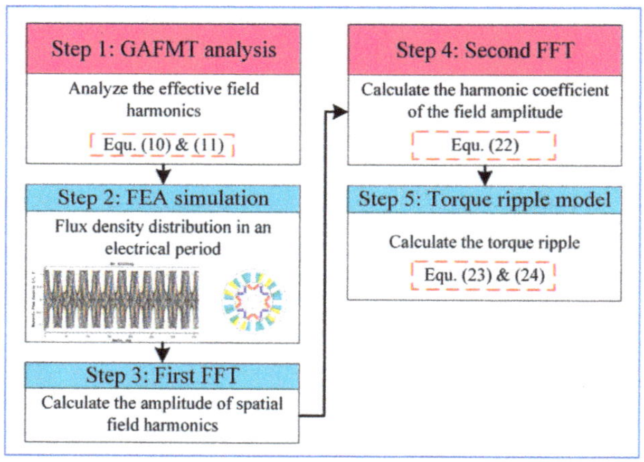

Figure 6. Flow chart of the proposed method.

In order to simplify the calculation, only the DC component, first order harmonic, and second order harmonic of the field harmonic amplitude are considered in the analyses. Suppose the amplitude of kth effective field harmonic can be nearly expressed as

$$A_k \approx C_{k0} + C_{k1}\cos(\gamma\omega t + \varphi_{k1}) + C_{k2}\cos(2\gamma\omega t + \varphi_{k2}) \tag{22}$$

where A_k denotes the field harmonic amplitude. C_{k0} denotes the DC component of field harmonic amplitude. C_{k1}/φ_{k1} and C_{k2}/φ_{k2} denote the first and second order harmonic amplitudes and phases, respectively, and γ denotes the least common multiple (LCM) of the stator and rotor poles.

4.2. Cogging Torque Model

Substituting Equation (22) into Equation (21), and neglecting the small amplitude components, the cogging torque model can be expressed as

$$T_{cog} \approx \frac{\pi(R_s^2 - R_r^2)l_{stk}}{4\mu_0}\sum_k\{2C_{k0}C_{k1}\gamma\sin(\gamma\omega t + \varphi_{k1}) + 8C_{k0}C_{k2}\gamma\sin(2\gamma\omega t + \varphi_{k1})\} \tag{23}$$

4.3. Modulation Torque Ripple Model

Similarly, to simplify the calculation, only the DC component of field harmonic amplitude is considered in analyzing the modulation torque ripple. Suppose the DC components of the field harmonic amplitude and the temporal orders of the field harmonic with the pole pairs of z are defined as C_{zm}/m and C_{za}/a, respectively, for the PM-excited and armature fields. Then, the modulation torque ripple can be expressed as

$$T_{modu} \approx \frac{\pi(R_s^2 - R_r^2)l_{stk}}{2\mu_0}\sum_z\{C_{zm}C_{za}(m+a)p\sin((m+a)p\omega t)\} \tag{24}$$

5. Demonstration of Torque Ripple Model

To verify the proposed torque ripple model, FEA-based simulations are carried out. The prototype topology is shown in Figure 7, and the key parameters are given in Table 2.

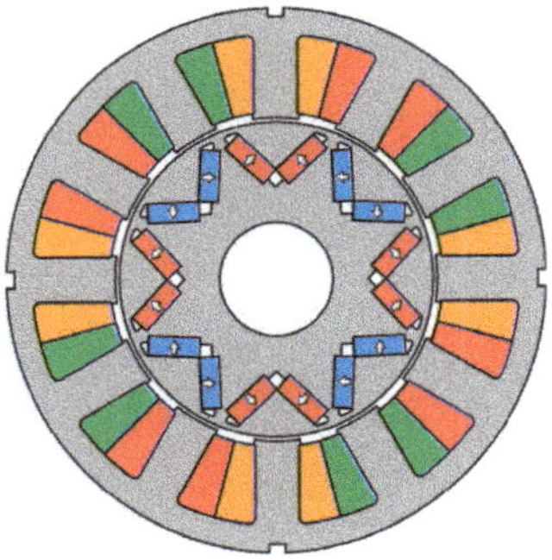

Figure 7. Prototype topology of a 12/8 IPM machine.

Table 2. The key parameters of the prototype machine.

Parameters	Value	Units	Parameters	Value	Units
Outer diameter of the stator	100	mm	Thickness of the PM	3.6	mm
Inner diameter of the stator	59.6	mm	Stack length	75	mm
Radian of the stator tooth	8.1	mm	Nominal torque	10	Nm
Radian of the stator slot	7	mm	Angle of the V-Shape PMs	95	deg
Thickness of the magnetic bridge	0.7	mm	Outer diameter of the rotor	58.1	mm

5.1. PM-Excited Cogging Torque

To verify the PM-excited torque ripple, FEA simulations under no-load condition are carried out. Figure 8 shows the characteristics of the modulated PM-excited field harmonics. Figure 8a,b show the waveform and the spatial spectrum of the PM-excited flux density in the radial direction. It can be seen that there are multiple harmonics in the airgap, where the harmonics with the pole pairs of 4, 12, and 20 are mainly generated by primitive modulation, and those of 8 and 16 are generated by summation and differential modulations. The rotation speed of harmonics with pole pairs of 12 is three times as fast as that of the harmonics with pole pairs of 4, 8, 16. All the above results are consistent with Equation (10).

In conjunction with Figure 8b,c, the harmonic amplitudes with the pole pairs of 16 and 8 may not be equal at a certain time, but their average values are equal in an electrical period. Obviously, this phenomenon demonstrates not only the correctness of GAFMT analysis but also the effect of stator–rotor relative position changes on the field harmonics amplitude.

As shown in Figure 8c, the field harmonics amplitude contains AC components, and the frequency is related to the LCM of the stator and rotor poles. Using the dual FFT method proposed in Section 4, the field harmonics amplitude and the parameters needed in Equation (22) can be obtained. Then, the PM-excited torque ripple can be calculated, as shown in Figure 9.

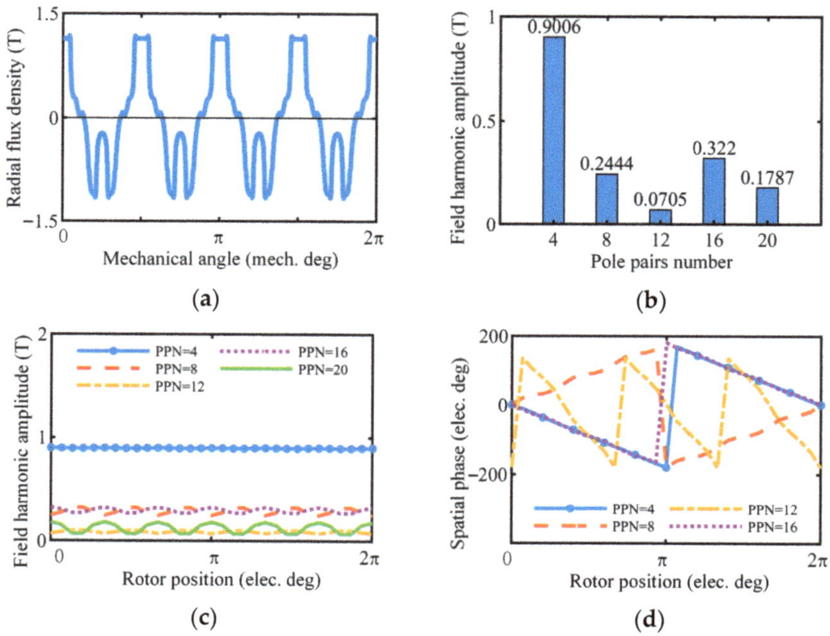

Figure 8. Modulated PM-excited field harmonics. (**a**) Distribution. (**b**) Spatial spectrum. (**c**) Harmonic amplitudes in one electrical period. (**d**) Harmonic phases in one electrical period.

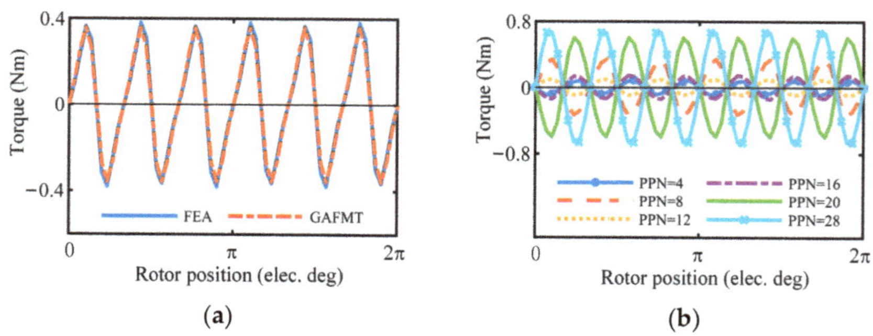

Figure 9. PM-excited cogging torque. (**a**). Results comparison from GAFMT and FEA. (**b**). Contributions of different PM-excited field harmonics.

As shown in Figure 9a, the GAFMT-based PM-excited cogging torque coincides well with the FEA-based result, and the error is mainly due to the neglect of the small value components during calculation. As shown in Figure 9b, even though the torque ripples contributed by harmonics with pole pairs of 20 and 28 are quite high, their phase is 180° apart, so their synthetic torque ripple is considerably reduced, and the same is true for harmonics with pole pairs of 4 and 12.

It is therefore difficult to say that the PM-excited cogging torque is caused by one or several harmonics. With the help of the dual FFT analysis method, it can be seen that the amplitude of all harmonics varies with the rotor position, and they all produce torque ripple. This may mean that the effect of all harmonics on torque ripple should be taken into account when optimizing the machine structure.

5.2. Armature Cogging Torque

The analysis method for armature-excited cogging torque is similar to that of PM-excited cogging torque. One factor that should be noted is that FEA simulations are carried out by using the frozen permeability method to reflect the effect of magnetic bridge saturation. The armature current amplitude is 5 A, and the phase is 0°.

The harmonic with pole pairs of 12 is generated by summation modulation from the harmonic with pole pairs of 4; its rotation speed is three times as fast as that of the harmonic with pole pairs of 4, and its appearance can be explained by the GAFMT, as expressed in Equation (11). FEA simulations also prove this, as shown in Figure 10d.

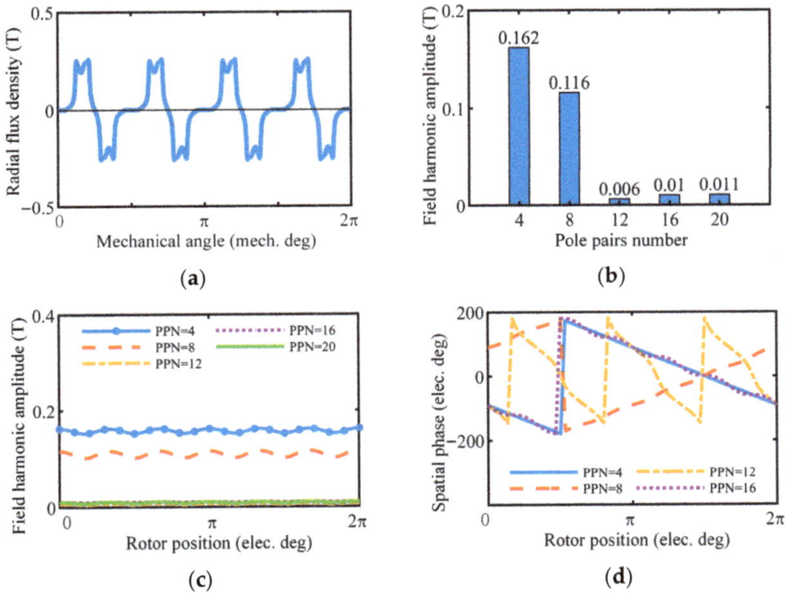

Figure 10. Modulated armature field harmonics. (**a**) Distribution. (**b**) Spatial spectrum. (**c**) Harmonic field amplitude in one electrical period. (**d**) Harmonic field phase in one electrical period.

Similar to the PM-excited field, the amplitude of the armature field harmonics is rippled due to the asynchronous modulation of the rotor salient pole reluctance, as shown in Figure 10c. The field harmonics amplitude is analyzed, and the armature-excited cogging torque is calculated by Equation (20); the results are shown in Figure 11. Compared to other field harmonics, the amplitudes of harmonics with pole pairs of 4 and 8 are high and cause the main armature-excited cogging torque.

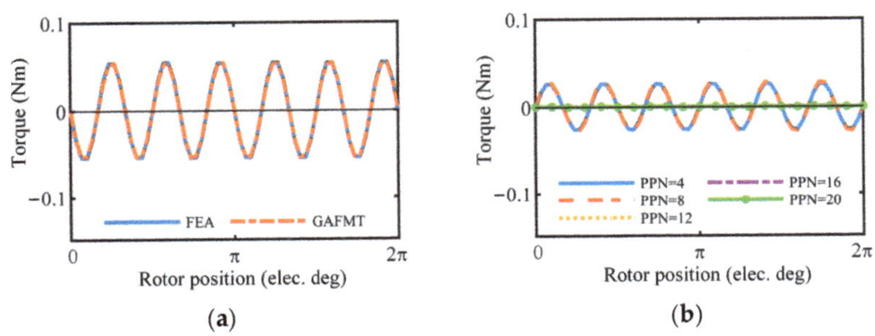

Figure 11. Armature-excited cogging torque. (**a**) Comparison of GAFMT result and FEA simulation result. (**b**) Armature-excited cogging torque produced by different field harmonics.

5.3. Modulation Torque Ripple

As analyzed above, the modulation torque ripple is produced by the difference in the rotation speed of the PM-excited field and armature field. For the prototype, the magnetic field distribution and rotation speed are analyzed by GAFMT and given in Table 3.

Table 3. Parameters of PM-excited and armature field.

PM-Excited Field		Armature Field	
Pole pairs	Rotation speed	Pole pairs	Rotation speed
20 28	20ω 20ω	20 28	-4ω 4ω
Pole pairs	C_{zm}	Pole pairs	C_{za}
20 28	0.1223 0.1075	20 28	0.0093 0.0266

Substituting the machine parameters into Equations (10) and (11), the rotation speed of PM-excited harmonics with pole pairs of 20 and 28 is 20ω and 28ω, respectively, while that of the armature-excited harmonics with the same pole pairs is $\pm 4\omega$, respectively. Then, substitute these results into Equation (24). The frequency of modulation torque ripple is 24ω, which is the same as the PM-excited cogging torque and the armature-excited cogging torque.

The values of C_{zm} and C_{za} in Equation (24) are also shown in Table 3. The modulation torque ripple curve is shown in Figure 12. Since the higher temporal harmonics are ignored to simplify the calculation, the high-frequency torque ripple of the GAFMT-based result is missing, but overall, the low-frequency value is relatively close, proving the validity of the proposed torque ripple model.

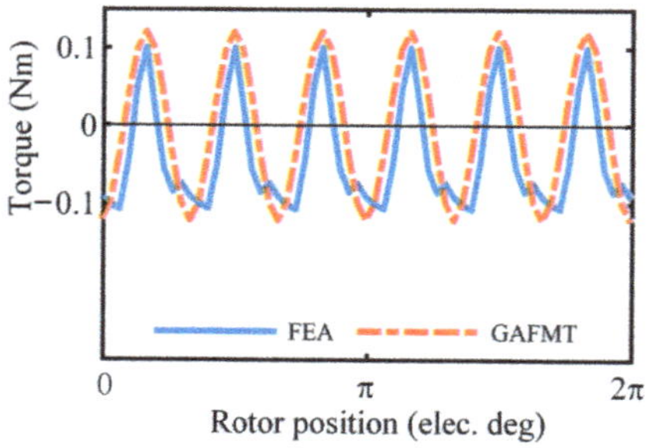

Figure 12. Modulation torque ripple.

6. Experimental Results

In order to verify the proposed torque ripple model, experiments are carried out. The experimental platform is shown in Figure 13, where the load machine is a brush dc machine, the exciting voltage of which is set as 80 V, and the armature winding is connected to a sliding resistor, as shown in the picture. A high-precision torque sensor is used to measure the torque waveforms, and the dSPACE 1103 platform is used to control the prototype.

Figure 14 is the PM-excited cogging torque measured by the sensor when the IPM machine is rotating at 150 rpm, and the amplitude of the fundamental torque ripple harmonic is 0.403 Nm. This result is in good agreement with the FEA simulation result, which is 0.38 Nm, as shown in Figure 9, and corresponds to the torque ripple model based on the GAFMT.

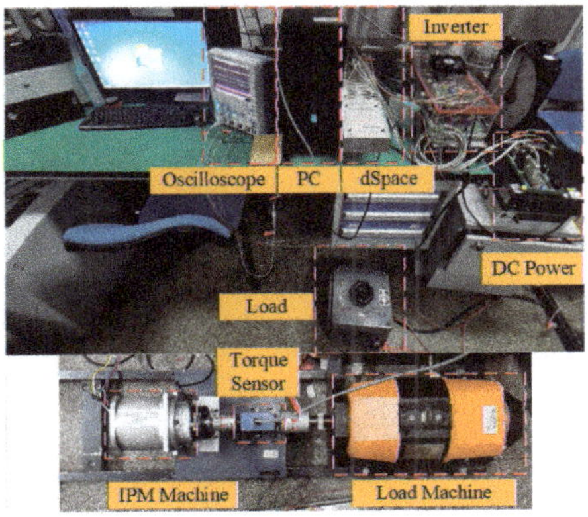

Figure 13. Experimental setup.

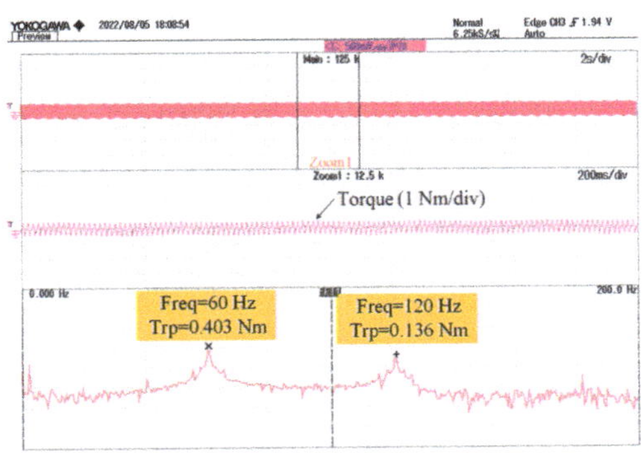

Figure 14. PM-excited cogging torque (150 rpm).

Figure 15a–c show the torque performance under different rotation speeds while the armature current is set as 1 p.u. (1 p.u. = 3.125 A). According to the proposed model, the torque ripple does not change along with the speed, but as shown in the figures, the torque ripple increases when the speed is 450 rpm. This phenomenon is caused by the low bandwidth of the controller and can be improved by parameter optimization. In order to reduce the influence of the controller parameters on the experimental results, the speed of the subsequent experiments will be set at a low speed.

As discussed in Section 4, the armature current related torque ripple component includes the armature-excited cogging torque and modulation torque ripple; when the amplitude of the armature current increases, these two torque ripples increase. However, all three torque ripple components have the same fundamental frequency; the total torque ripple is not only determined by their amplitude but also influenced by their phase. Figure 15a,d,e show the experimental results when the armature current amplitude is changed, while Figure 15f shows the comparison of the FEA results, the proposed model results, and the experimental results. When the armature current amplitude changes from 0.5 p.u. to 1.5 p.u., the torque ripple decreases from 0.414 Nm to 0.29 Nm in the FEA simulation results, 0.395 Nm to 0.337 Nm in the model results, and 0.428 Nm to 0.334 Nm in experimental results, respectively. It can be concluded that the proposed model can provide a numerically accurate result for the variation of torque ripple with the armature magnetic field, and it confirms the correctness of the three torque ripple components proposed by the GAFMT-based model.

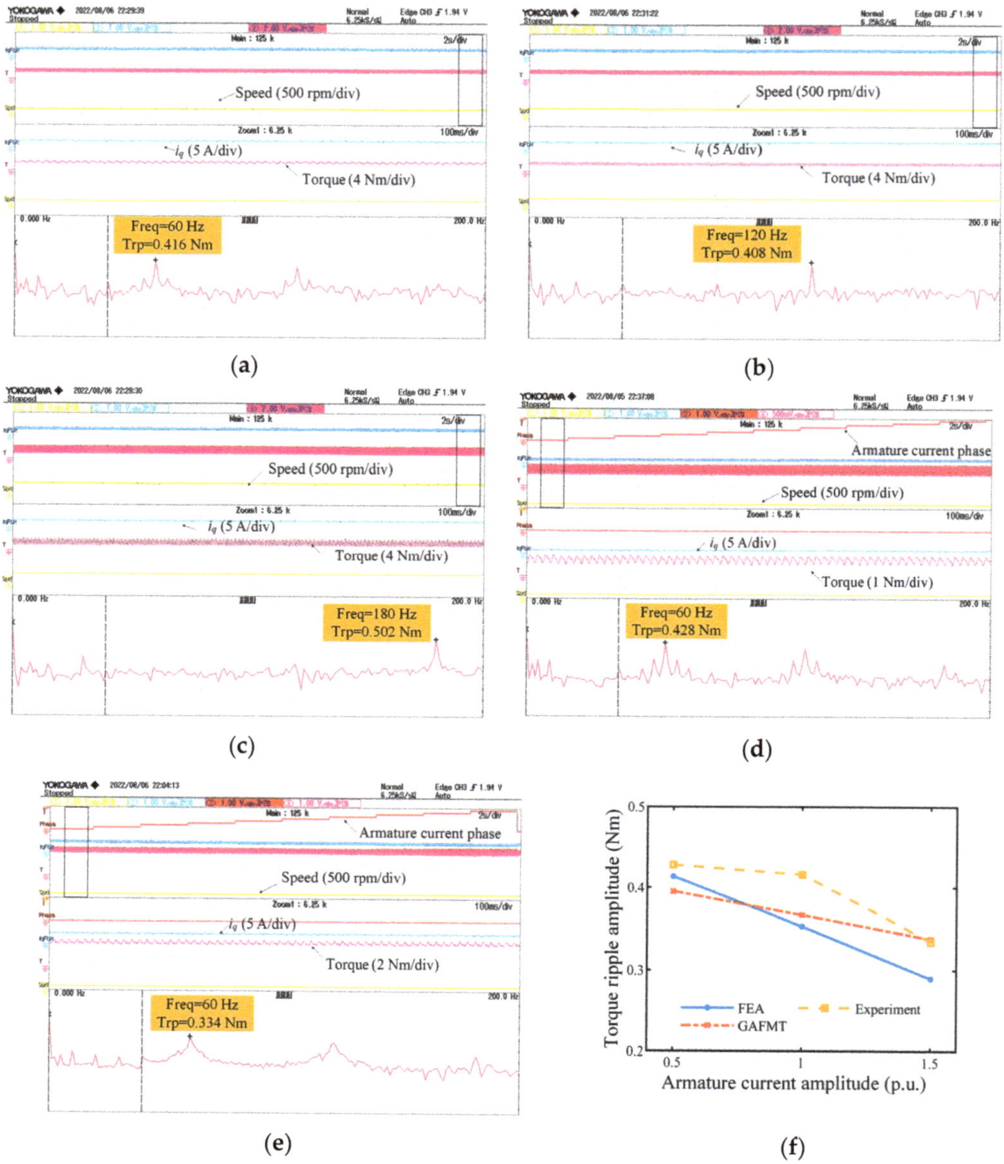

Figure 15. Experimental torque performance. (**a**). 150 rpm, i_q = 1 p.u., i_d = 0 A. (**b**). 300 rpm, i_q = 1 p.u., i_d = 0 A. (**c**). 450 rpm, i_q = 1 p.u., i_d = 0 A. (**d**). 150 rpm, i_q = 0.5 p.u., i_d = 0 A. (**e**). 150 rpm, i_q = 1.5 p.u., i_d = 0 A. (**f**). Comparison of torque ripple amplitudes under different armature current amplitudes (150 rpm).

7. Conclusions

In this paper, the spatial characteristics of both PM-excited and armature magnetic field harmonics were first analyzed by using the GAFMT. The influence of the magnetic bridge on the modulation effect was investigated, showing that both the rotor and the stator of the typical V-Shape IPM machine can be considered as the salient pole reluctance.

Then, a novel torque ripple model, which contains three components, is proposed. The total torque ripple is divided into PM-excited cogging torque, armature cogging torque, and modulation torque ripple. Thanks to the general nature of the GAFMT, the proposed model can be extended to any other machine. By subdividing the total torque ripple, it is possible to reveal the mechanisms of torque ripple in different operating conditions and provide guidance for structure or control method-based optimization under full operating conditions.

Furthermore, to achieve an accurate torque ripple model, the FEA-assisted dual FFT method was proposed to investigate the amplitude ripple of the magnetic field harmonics. As the results in the previous sections show, the magnetic field harmonics have, in reality, a significant space–time coupling characteristic and are traveling through waves whose amplitude varies periodically with the relative position between the stator and rotor.

Finally, FEA simulations and experiments were carried out to verify the correctness of the proposed torque ripple model.

On the basis of the proposed method in this paper, the future research is to further explore the potential of the GAFMT, improve the accuracy of the flux density calculation, and reduce the dependence of the model on the FEA results.

Author Contributions: Conceptualization, J.Z.; methodology, J.Z., M.C., W.Y. and W.H.; formal analysis, J.Z. and M.C.; investigation, M.C. and W.H.; writing—original draft preparation, J.Z. and M.C.; supervision, M.C. and W.H.; funding acquisition, M.C. All authors have read and agreed to the published version of the manuscript.

Funding: This research was funded by National Natural Science Foundation of China, grant number 51991381.

Data Availability Statement: Not applicable.

Conflicts of Interest: The authors declare no conflict of interest.

References

1. Liu, Q.; Hameyer, K. High-performance adaptive torque control for an IPMSM with real-time MTPA operation. *IEEE Trans. Energy Convers.* **2017**, *32*, 571–581. [CrossRef]
2. Liu, X.; Chen, H.; Zhao, J.; Belahcen, A. Research on the performances and parameters of interior PMSM used for electric vehicles. *IEEE Trans. Ind. Electron.* **2016**, *63*, 3533–3545. [CrossRef]
3. Hao, Y.; Li, J.; Bianchi, F.; Ciuti, G.; Dario, P.; Huang, Q. Analytical magnetic model towards compact design of magnetically-driven capsule robots. *IEEE Trans. Med. Rob. Bionics* **2020**, *2*, 188–195. [CrossRef]
4. Zhang, W.; Xu, Y.; Huang, Y.; Zou, J. Reduction of high-frequency vibration noise for dual-branch three-phase permanent magnet synchronous motors. *Chin. J. Elect. Eng.* **2020**, *6*, 42–51. [CrossRef]
5. Yamazaki, K.; Utsunomiya, K.; Ohiwa, H. Mechanism of torque ripple generation by time and space harmonic magnetic fields in permanent magnet synchronous motors. *IEEE Trans. Ind. Electron.* **2022**, *69*, 9884–9894. [CrossRef]
6. West, H.R. The cross-field theory of alternating-current machines. *Trans. AIEE* **1926**, *XLV*, 466–474.
7. Fiennes, J. New approach to general theory of electrical machines using magnetic equivalent circuits. *Proc. IEE.* **1973**, *120*, 94–104. [CrossRef]
8. Hansen, K.L. The rotating magnetic field theory of AC machines. *Trans. AIEE* **1925**, *XLIV*, 340–348.
9. Cheng, M.; Han, P.; Hua, W. General airgap field modulation theory for electrical machines. *IEEE Trans. Ind. Electron.* **2017**, *64*, 6063–6074. [CrossRef]
10. Cheng, M.; Han, P.; Du, Y.; Wen, H.; Li, X. A tutorial on general air-gap field modulation theory for electrical machines. *IEEE J. Emerg. Sel. Topics Power Electron.* **2022**, *10*, 1712–1732. [CrossRef]
11. Cheng, M.; Hua, W.; Zhang, J.; Zhao, W. Overview of stator-permanent magnet brushless machines. *IEEE Trans. Ind. Electron.* **2011**, *58*, 5087–5101. [CrossRef]
12. Toba, A.; Lipo, T.A. Novel dual-excitation permanent magnet vernier machine. In Proceedings of the IEEE IAS Annual Meeting, Pheonix, AZ, USA, 3–7 October 1999.
13. Zhu, X.; Hua, W.; Wang, W.; Huang, W. Analysis of back-EMF in flux-reversal permanent magnet machines by air gap field modulation theory. *IEEE Trans. Ind. Electron.* **2019**, *66*, 3344–3355. [CrossRef]
14. Wen, H.; Cheng, M. Unified analysis of induction machine and synchronous machine based on the general airgap field modulation theory. *IEEE Trans. Ind. Electron.* **2019**, *66*, 9205–9216. [CrossRef]
15. Zhou, J.; Cheng, M.; Wen, H.; Yan, X.; Tong, M.; Wang, W. Modeling and suppression of torque ripple in PMSM based on the general airgap field modulation theory. *IEEE Trans. Power Electron.* **2022**, *37*, 12502–12512. [CrossRef]
16. Cheng, M.; Wen, H.; Han, P.; Zhu, X. Analysis of airgap field modulation principle of simple salient poles. *IEEE Trans. Ind. Electron.* **2019**, *66*, 2628–2638. [CrossRef]

17. Shin, K.H.; Choi, J.Y.; Cho, H.W. Characteristic analysis of interior permanent magnet synchronous machine with fractional slot concentrated winding considering nonlinear magnetic saturation. *IEEE Trans. Appl. Supercond.* **2016**, *26*, 5200404. [CrossRef]
18. Wang, P.; Hua, W.; Zhang, G.; Wang, B.; Cheng, M. Principle of flux-switching permanent magnet machine by magnetic field modulation theory part I: Back-electromotive-force generation. *IEEE Trans. Ind. Electron.* **2022**, *69*, 2370–2379. [CrossRef]

energies

MDPI

Article

Parameter-Free Model Predictive Current Control for PMSM Based on Current Variation Estimation without Position Sensor

Laiwu Luo, Feng Yu *, Lei Ren and Cheng Lu

School of Electrical Engineering, Nantong University, Nantong 226019, China; luolaiwu@ntu.edu.cn (L.L.); ee_renlei@ntu.edu.cn (L.R.); lucheng@ntu.edu.cn (C.L.)
* Correspondence: yufeng628@ntu.edu.cn

Abstract: To remove parameter dependence in existing sensorless control strategies, a parameter-free model predictive current control is proposed for permanent magnet synchronous motor without any position sensor. First, the current variation during one sampling period is analyzed and divided into two elements: natural attenuation and forced response. Second, recursive least squares algorithm is utilized to estimate the future current variation so that the model predictive current control can be successfully executed paying no attention to motor parameters. Meanwhile, the position information is obtained by the arc tangent function according to the estimated forced response of current variation. At last, experimental results verify that the estimation errors of rotor position are reduced to around 0.1 rad with smaller current prediction error even at low speed where no motor parameters are required.

Keywords: sensorless control; parameter-free; model predictive current control; permanent magnet synchronous motor; recursive least squares algorithm

Citation: Luo, L.; Yu, F.; Ren, L.; Lu, C. Parameter-Free Model Predictive Current Control for PMSM Based on Current Variation Estimation without Position Sensor. *Energies* **2023**, *16*, 6792. https://doi.org/10.3390/en16196792

Academic Editors: Jung Rae Kim and Lorand Szabo

Received: 23 August 2023
Revised: 14 September 2023
Accepted: 20 September 2023
Published: 24 September 2023

1. Introduction

Permanent magnet synchronous motor (PMSM) is identified as a high-performance exchange motor. It has played an irreplaceable role in industrial and civil production [1]. Nowadays, its control technology is always evolving owing to the advancement of motion control theory, power electronics technology and sensor design. Model predictive control (MPC) strategy is viewed as a promising way to replace the PI modem, directly taking into account the nonlinearity and constraint characteristics of the system [2,3]. Depending on the operation principle, MPC can be classified into two categories, namely the continuous control set MPC (CCS-MPC) and the finite control set MPC (FCS-MPC). Specifically, the FCS-MPC strategy directly tracks the reference states through the enumeration of possible switching states, so that the applied switching state for inverter can be decided simply [4,5]. In this manner, the external modulation stages are avoided, which is an indispensable part of CCS-MPC [6]. At present, model predictive current control (MPCC) has been widely used with the aim to predict future currents.

In practical applications, high-precision sensors increase the design cost of the control system, and the mechanical encoders desire additional assembly space and system maintenance. In addition, the stability of communication between the encoder and the motor drive is inevitably affected by external signals and noise. Moreover, when the sensors are installed in a PMSM with low moment of inertia, rotational inertia of rotor increases incurring the deteriorative robustness [7–9]. In addition, in some more complex operating environments, the motor drive system ensures the safety and stability of the system in addition to ensuring high dynamic responsiveness and gauge stability. Position sensors are a common part of the motor control system failure, and in some cases can be replaced by sensorless control technology [10].

In principle, the positionless sensing program can be generally divided into two categories: one is a program based on the electromagnetic variation relationship in a mathematical model, and the other is a position estimation strategy based on a PMSM emphasis effect. The former can be subdivided into an open-circle calculation program based upon an ideal model and a closed-circuit calculation method based on an observer [11], while the latter is typically based on high-frequency injection methods based on non-ideal properties achieved [12]. Many sensorless control methods are in conjunction with MPCC. Refs. [13–15] put effort into high-frequency information injection to identify the rotor speed, which takes advantage of unequal quadrature- and direct-axis inductance caused by inductance magnetic saturation effect. As a whole, high-frequency injection methods have some problems such as poor dynamic performance and reduced bandwidth. Also, the performance of applied power electronic device is critical.

Instead, refs. [16–20] advocate for the design of observers that function to estimated rotor position. In [16], an adaptive full-order observer is adopted to smoothly start a free-running motor without any position sensors. The feedback gain matrix is conceived to realize that the estimated speed can trace the actual value even if the initial difference is large. On the contrary, the rotor position is estimated via a disturbance observer in [17], and the estimation error is effectively reduced by considering the iron loss. Even in the quadrotor system, observers like finite-time extended state observer and adaptive neural network observer are adopted to estimate the position information [18]. In addition, the sliding mode observer (SMO) and the full-order state observer are utilized in the sensorless control of [19,20], respectively. Nonetheless, it is worth mentioning that these sensorless control methods are dependent on the motor parameters, which fail to remain constant in all conditions owing to the variable operation environment [21,22]. Thus, the strategies with strong parameter robustness attract a lot of attention to keep the normal operation when parameters are mismatched [23].

Reference [24] analyzes the influence of parameter mismatch on the proposed sensorless method. The injected high-frequency voltage is viewed as the reference value and added into the cost function for voltage tracking. Therefore, the robustness against parameter mismatch is improved without any complicated work. On the premise of parameter-free MPCC based on the ultra-local model, ref. [25] builds a sliding mode observer for a sensorless driving system, and the prediction horizon can be adjusted online considering prediction error. An improved-position self-detecting observer is adopted in [26], where the estimation accuracy of rotor position is improved. Meanwhile, the parameter-adaptive SMO is devoted to suppressing the influence of motor parameter variation. It is obvious that most sensorless methods achieve parameter-free control owing to the observers. However, the application of observers indicates that the mathematical model of the motor is still indispensable in the control. For example, the ultralocal model in [24] predicts the future current by virtue of inductance parameters. Moreover, when more observers participate in the control, the increased complexity of the system is accompanied by heavy tuning work.

To completely get rid of the dependence of motor parameters in position-free control, the current variations corresponding to active voltage vectors are dissected in this paper, based on which the prediction of future current and rotor position estimation are achieved via the current variations obtained from the recursive least squares (RLS) algorithm. The contributions of this paper are summarized as follows:

1. Taking the current variations as the estimation target, a parameter-free MPCC is designed on the basis of RLS, where the natural attenuation and forced response of current variations are estimated accurately. It successfully avoids the effects of lost initial data in [25] and the problem of current variation renewal stagnates in [23].
2. To estimate the rotor position through the known current variations, the forced response value is analyzed together with the active voltage vector, after which an accurate rotor position angle can be obtained by the built arc tangent function. Parameter dependence in [24,26] is successfully overcome.

3. Both simulated and experimental results verify the correctness and effectiveness of the proposed method.

The remainder of article is organized as follows. The model of the PMSM drive system is provided in Section 2, where the current variation is analyzed carefully. In Section 3, the proposed sensorless parameter-free MPCC is revealed in terms of future current prediction and rotor position estimation. Afterwards, the experimental results verify the effectiveness and superiority of the proposed control strategy in Section 4. Finally, this study is concluded in Section 5.

2. Model of PMSM Drive System

2.1. Mathematical Model of Motor

For a convenient design of a controller, the mathematical model is usually built in synchronous rotating coordinates, so that the stator voltage of the concerned three-phase PMSM [24] can be obtained as

$$\begin{bmatrix} u_d \\ u_q \end{bmatrix} = \begin{bmatrix} R & -\omega_e L_q \\ \omega_e L_d & R \end{bmatrix} \begin{bmatrix} i_d \\ i_q \end{bmatrix} + \begin{bmatrix} L_d & 0 \\ 0 & L_q \end{bmatrix} \frac{\mathrm{d}}{\mathrm{d}t} \begin{bmatrix} i_d \\ i_q \end{bmatrix} + \begin{bmatrix} 0 \\ \omega_e \psi_f \end{bmatrix}, \tag{1}$$

where u, i and L refer to the stator voltage, current and inductance, respectively. Subscripts d and q represent the corresponding components of these parameters in dq-axes. R, ω_e and ψ_f denote the stator resistance, electrical angular velocity and permanent magnet flux linkage, respectively. In this paper, PMSM is driven by a traditional two-level inverter, which can produce eight voltage vectors as exhibited in Figure 1.

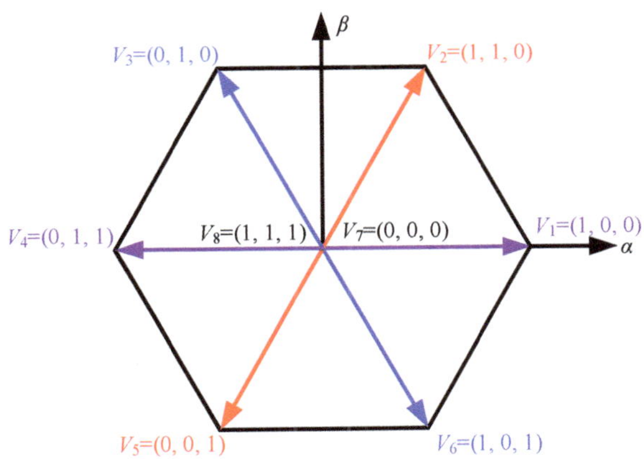

Figure 1. Space voltage vector distribution in a two-level inverter.

2.2. Model of Current Variations

By rearranging the stator voltage Equation (1), the differentials of dq-axes currents can be calculated as

$$\frac{\mathrm{d}}{\mathrm{d}t} \begin{bmatrix} i_d \\ i_q \end{bmatrix} = \begin{bmatrix} -R/L_d & \omega_e L_q/L_d \\ -\omega_e L_d/L_q & -R/L_q \end{bmatrix} \begin{bmatrix} i_d \\ i_q \end{bmatrix} + \begin{bmatrix} u_d/L_d \\ u_q - \omega_e \psi_f/L_q \end{bmatrix}. \tag{2}$$

Relying on the Forward Euler method, dq-axes currents in $(k+1)$th are predicted as

$$\begin{bmatrix} i_d(k+1) \\ i_q(k+1) \end{bmatrix} = \begin{bmatrix} \frac{L_d - RT_s}{L_d} & \frac{L_q \omega_e T_s}{L_d} \\ \frac{L_q - RT_s}{L_q} & -\frac{L_d \omega_e T_s}{L_q} \end{bmatrix} \begin{bmatrix} i_d(k) \\ i_q(k) \end{bmatrix} + \begin{bmatrix} \frac{T_s}{L_d} u_d(k) \\ \frac{T_s}{L_q} u_q(k) - \frac{\psi_f \omega_e T_s}{L_q} \end{bmatrix}. \tag{3}$$

Therefore, the current variations between two consecutive sampling periods can be expressed as

$$\begin{bmatrix} \Delta i_d | V_j \\ \Delta i_q | V_j \end{bmatrix} = \underbrace{\begin{bmatrix} -RT_s/L_d & \omega_e L_q T_s/L_d \\ -\omega_e L_d T_s/L_q & -RT_s/L_q \end{bmatrix} \begin{bmatrix} i_d \\ i_q \end{bmatrix} + \begin{bmatrix} 0 \\ -\omega_e \psi_f T_s/L_q \end{bmatrix}}_{\delta i_{x0}} + \underbrace{\begin{bmatrix} (u_d|V_j)T_s/L_d \\ (u_q|V_j)T_s/L_q \end{bmatrix}}_{\delta i_x|V_j}, \quad (4)$$

where $| V_j$ refers to parameters corresponding to voltage vector V_j, ($j \in \{1, 2, \ldots, 8\}$).

Obviously, the current variation can be viewed as the sum of two parts, namely natural attenuation δi_{x0} and forced response $\delta i_x | V_j$ ($x = d, q$). The former part depends on the stator currents and motor speed. The motor speed can be thought of as a constant value in a few time steps since the electrical time constant is substantially smaller than the mechanical time constant. The stator current is also essentially constant in a steady state. As a result, it is possible to think of the natural attenuation as a constant value that corresponds to the zero-voltage vectors. On the other hand, the latter part is merely decided by applied active voltage vectors. That is to say, the forced response of current variation is closely associated with the amplitude and position of the voltage vectors.

3. Proposed Sensorless Parameter-Free MPCC Strategy

3.1. Parameter-Free MPCC

Due to the incredibly brief sample time of the microprocessor, the current in each sampling period may generally be approximated as a linear variation. By sensing the stator current at the start and end of each sampling period, it is possible to calculate the current variations as

$$\begin{bmatrix} \Delta i_d | V_j \\ \Delta i_q | V_j \end{bmatrix} = \begin{bmatrix} i_d(k,1)|V_j - i_d(k,2)|V_j \\ i_q(k,1)|V_j - i_q(k,2)|V_j \end{bmatrix}, \quad (5)$$

where Δi represents the stator current variations. $i(k,1)$ and $i(k,2)$ represent the stator current measured at the beginning and the end of the (k)th, respectively.

It should be noted that, in accordance with (5), the current detections are conducted twice throughout each sampling interval. In addition, each control period must begin and end with a sudden voltage signal, suggesting that the current may not be accurately identified. In pursuit of precise current values, it is imperative to conduct dual detections, namely post-commencement and pre-conclusion, during each sampling period. Nevertheless, such an approach consequently imposes a substantial computational burden on microprocessors, rendering real-time implementation prohibitively challenging. To address this dilemma, an alternative procedure advocates the detection of singular currents at each sampling period. This strategy is predicated on the assumption that the value of the stator current at the onset of a given sampling period is equal to the value at the culmination of the preceding sampling period.

The schematic diagram of primary current sampling is illustrated in Figure 2, where (k − 1) and (k + 1) denote the system parameters at the (k − 1)th and (k + 1)th period, respectively. Notably, a brief interval of Δt_0 exists between the current detection and the conclusion of the sampling period, which is shorter than one twentieth of the sampling period. The purpose of this interval is to mitigate the impact of abrupt voltage fluctuations on the precision of current measurement.

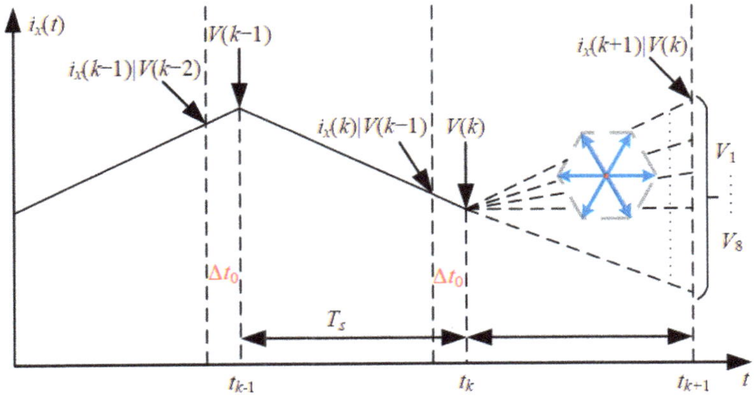

Figure 2. Schematic diagram of the stator current prediction based on once current detection.

In accordance with the depicted stator current prediction methodology illustrated in Figure 2, the current variation at the (k)th instance can be derived as

$$\begin{bmatrix} \Delta i_d|V(k-1) \\ \Delta i_q|V(k-1) \end{bmatrix} = \begin{bmatrix} i_d(k)|V(k-1) - i_d(k-1)|V(k-2) \\ i_q(k)|V(k-1) - i_q(k-1)|V(k-2) \end{bmatrix}. \tag{6}$$

Analogously, the variation value of the stator current at the $(k+1)$th can be formulated as

$$\begin{bmatrix} \Delta i_d|V(k) \\ \Delta i_q|V(k) \end{bmatrix} = \begin{bmatrix} i_d(k+1)|V(k) - i_d(k)|V(k-1) \\ i_q(k+1)|V(k) - i_q(k)|V(k-1) \end{bmatrix}. \tag{7}$$

It is imperative to acknowledge that the computations of the current variations $\Delta i_d \mid V(k)$ and $\Delta i_q \mid V(k)$, as described in Equation (8), are not feasible due to the unavailability of measured stator currents $i_d(k+1) \mid V(k)$ and $i_q(k+1) \mid V(k)$ during the (k)th sampling interval. However, providentially, the brevity of the sampling period *Ts* permits the insignificance of changes in current variations associated with the identical voltage vector. Stated differently, $\Delta i_d \mid V(k)$ and $\Delta i_q \mid V(k)$ can be substituted with their preceding values, namely

$$\Delta i_x|V(k) \approx \Delta i_{x,pre}|V(k), V(k) \in \{V_1, V_2, \ldots, V_8\}, \tag{8}$$

where subscript *pre* denotes the antecedent state, which can be securely retained within the microprocessor controller.

By amalgamating Equations (8) and (9), the *dq*-axes current at the $(k+1)$th can be predicted:

$$\begin{bmatrix} i_d(k+1)|V_j \\ i_q(k+1)|V_j \end{bmatrix} = \begin{bmatrix} i_d(k) + \Delta i_{d,pre}|V_j \\ i_q(k) + \Delta i_{q,pre}|V_j \end{bmatrix}. \tag{9}$$

The cost function of MPCC can be formulated in the following, taking into account minimum current tracking error. Subsequently, the corresponding optimal voltage vector can be chosen for the inverter.

$$g_j = \left| i_d^{\text{ref}}(k+1) - i_d(k+1)\middle|V_j \right| + \left| i_q^{\text{ref}}(k+1) - i_q(k+1)\middle|V_j \right|. \tag{10}$$

3.2. Estimation of Current Variation

According to Section 2.2, the current variations can be analyzed as

$$\Delta i_x(k) = \delta i_{x0}(i_d, i_q, \omega_e) + \delta i_x(i_d, i_q, \theta_e, n), \tag{11}$$

where the forced response is further inferred as

$$\delta i_x(i_d, i_q, \theta_e, n) = \frac{2T_s U_{dc}}{3L_x}\cos\left((V-1)\frac{\pi}{3} - \theta_e\right). \tag{12}$$

θ_e represents the electrical angle. V is the index of the applied voltage vector and $V \in \{1, 2, 3, 4, 5, 6\}$.

By the given

$$\begin{bmatrix} p_{1,x} \\ p_{2,x} \end{bmatrix} = \begin{bmatrix} \delta i_{x0}(i_d, i_q, \omega_e) \\ 2U_{dc}T_s/3L_x \end{bmatrix}. \tag{13}$$

dq-axis currents are rewritten as

$$\Delta i_x^v = \begin{bmatrix} 1 & \cos\left((n-1)\frac{\pi}{3} - \theta_e\right) \end{bmatrix} \begin{bmatrix} p_{1,x} & p_{2,x} \end{bmatrix}^{\mathrm{T}} = \boldsymbol{\psi}_x \boldsymbol{P}_d. \tag{14}$$

Thereafter, the RLS algorithm is considered to estimate the natural decay and forced response part of the current variations by virtue of the applied voltage vectors.

Input matrix $\boldsymbol{\psi}_x$ comprises the information of angular velocity, and output $\boldsymbol{y}_x$ is the actual current variations, based on which the system model is continuously trained in every update iteration. Finally, estimation matrix $\boldsymbol{P}_x$ containing the natural attenuation and forced response of current variations can be achieved. It is worth noting that the estimation of disparate parameters from closely resembling sampled currents poses a formidable challenge when one voltage vector is applied over two or more consecutive sampling periods. To this end, it is necessary to capture and store the voltage vector applied at each moment and determine the time at which the voltage vector starts that is different from the voltage vector applied at the beginning of $(k-1)$, namely $(k-m)$, to obtain the output value for system training. Consequently, the corresponding values pertaining to module input and output at their respective instances can be earmarked as

$$\boldsymbol{\psi}_x = \begin{bmatrix} \varphi_d \\ \varphi_q \end{bmatrix} = \begin{bmatrix} 1 & \cos\left((n-1)\frac{\pi}{3} - \theta_e\right) \\ 1 & \sin\left((n-1)\frac{\pi}{3} - \theta_e\right) \end{bmatrix}, \tag{15}$$

$$\begin{aligned} \mathbf{y_x}(k) &= \begin{bmatrix} \Delta i_x(k) & \Delta i_x(k-\mathrm{m}) \end{bmatrix} \\ &= \begin{bmatrix} i_x(k) - i_x(k-1) & i_x(k-m) - i_x(k-m-1) \end{bmatrix} \end{aligned}. \tag{16}$$

In this context, n serves as the index denoting the sequence of voltage vectors ($n = 1, 2, \ldots 8$). The magnitude of m corresponds to the count of temporal intervals from the inception of the voltage vector distinct from the preceding application at $(k-1)$. Detailed elucidation on the acquisition of electrical angle information is presented in the subsequent section.

Take the d-axis, for example; the initial valuations of the covariance matrix $\boldsymbol{Q_0}$ are given as

$$\boldsymbol{Q_0} = 10^6 \begin{bmatrix} 1 & 0 \\ 0 & 1 \end{bmatrix}. \tag{17}$$

Also, the gain matrix $\boldsymbol{G_0}$ is set as

$$\boldsymbol{G_0} = \boldsymbol{Q_0}\boldsymbol{\psi_{d0}}^{\mathrm{T}}\left(\boldsymbol{\psi_{d0}}\boldsymbol{Q_0}\boldsymbol{\psi_{d0}}^{\mathrm{T}} + \mu \boldsymbol{I}\right)^{-1}, \tag{18}$$

where $\boldsymbol{\psi_{d0}}$ denotes the initial estimation of angular velocity at the current moment, acquired through computations derived from (15), as $\boldsymbol{\psi_{d0}} = \begin{bmatrix} \varphi_d(k) \\ \varphi_d(k-m) \end{bmatrix}$; μ signifies the decaying factor, reflecting its propensity to gradually diminish over time; $\boldsymbol{I}$ presents an identity matrix with 2×2 dimensions.

By incorporating the initially determined values of covariance matrix $\boldsymbol{Q_x}$ and gain matrix $\boldsymbol{G_x}$ into Equations (19)–(21), a progressive calculation ensues. Following the acquisition of variable $\boldsymbol{P_d}$ in each iterative step, the evaluation comes into play by computing E_d

in accordance with Equation (22). Upon the fulfillment of the condition $E_d \leq 0.00001$, the ultimate estimation of the disparity in the d-axis current is produced as the final outcome, like $\boldsymbol{P_d} = \begin{bmatrix} p_{d1} \\ p_{d2} \end{bmatrix}$.

$$\boldsymbol{G_d}(k) = \boldsymbol{Q_d}(k-1)\boldsymbol{\psi_d}^{\mathrm{T}}(k)\left(\boldsymbol{\psi_d}\boldsymbol{Q_d}(k-1)\boldsymbol{\psi_d}^{\mathrm{T}}(k) + \mu \boldsymbol{I}\right)^{-1}, \tag{19}$$

$$\boldsymbol{Q_d}(k) = \frac{\boldsymbol{Q_d}(k-1) - \boldsymbol{G_d}(k)\boldsymbol{\psi_d}(k)\boldsymbol{Q_d}(k-1)}{\mu}, \tag{20}$$

$$\boldsymbol{P_d}(k) = \boldsymbol{P_d}(k-1) + \boldsymbol{G_d}(k)(\boldsymbol{y_d}(k) - \boldsymbol{\psi_d}(k)\boldsymbol{P_d}(k-1)), \tag{21}$$

$$E_d = \frac{\boldsymbol{\varphi_d}(k)\boldsymbol{P_d}(k) - \Delta i_d(k)}{\Delta i_d(k)}. \tag{22}$$

In a similar vein, through adherence to the parallel matrix iteration procedure, it becomes feasible to derive an estimation of the dissimilarity in the q-axis current, namely $\boldsymbol{P_q} = \begin{bmatrix} p_{q1} \\ p_{q2} \end{bmatrix}$. Thus, the current variations to the future time can be predicted as

$$\Delta i_d(k+1) = \boldsymbol{\varphi_d}(k)\boldsymbol{P_d}(k) = \begin{bmatrix} 1 & \cos\left((n-1)\frac{\pi}{3} - \theta_e\right) \end{bmatrix} \begin{bmatrix} p_{d1} \\ p_{d2} \end{bmatrix}, \tag{23}$$

$$\Delta i_q(k+1) = \boldsymbol{\varphi_q}(k)\boldsymbol{P_q}(k) = \begin{bmatrix} 1 & \sin\left((n-1)\frac{\pi}{3} - \theta_e\right) \end{bmatrix} \begin{bmatrix} p_{q1} \\ p_{q2} \end{bmatrix}. \tag{24}$$

Furthermore, the predictive value of future current states is given by

$$i_x(k+1) = i_x(k) + \Delta i_x(k+1). \tag{25}$$

3.3. Rotor Position Estimation

As illustrated in Figure 3, the compelled response value ensconced within the current differential flawlessly aligns itself with the spatial orientation of the applied active voltage vector. Meanwhile, the vector position is fixed in a two-phase rest coordinate system as shown in Figure 1. At the kth period, the optimal applied voltage vector selected by the value function shown in (10) is recorded, based on which angle γ between the applied vector and the α-axis in the two-phase rest coordinate system can be determined. The precise positional angles corresponding to individual voltage vectors are meticulously delineated in Table 1.

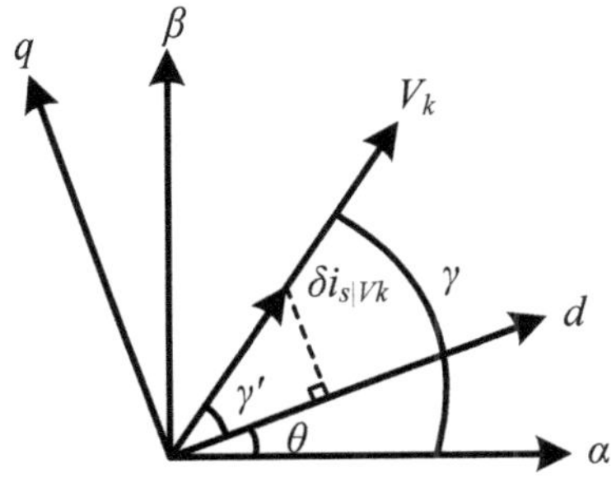

Figure 3. Schematic diagram of rotor position estimation.

Table 1. Voltage position definition.

Voltage Vector	S_{abc}	Vector Position Angle γ	Voltage Vector	S_{abc}	Vector Position Angle γ
V_1	100	0	V_2	110	$\pi/3$
V_3	010	$2\pi/3$	V_4	011	π
V_5	001	$4\pi/3$	V_6	101	$5\pi/3$
V_7	000	0	V_8	111	0

Conversely, through the application of the arctan function upon the forced response of dq-axes current variations relying on RLS, precise angle γ' between the applied voltage vector and the d-axis within the rotational coordinate system can be deduced. This distinctive interpretation of the relationship is portrayed by

$$\gamma\prime = \arctan\frac{\delta i_{q|S_j}}{\delta i_{d|S_j}}. \tag{26}$$

Therefore, the rotor position angle can be estimated by taking the difference between the γ and γ', and it is expressed as

$$\theta = |\gamma - \gamma'|. \tag{27}$$

Undeniably, this sensorless control strategy showcases remarkable parameter robustness by obviating the reliance on motor parameters in not only prospective current prediction but also rotor position estimation.

3.4. Summary

With reference to the analysis above, the following steps are required to implement the proposed method. And the structure of the proposed MPCC strategy is presented in Figure 4.

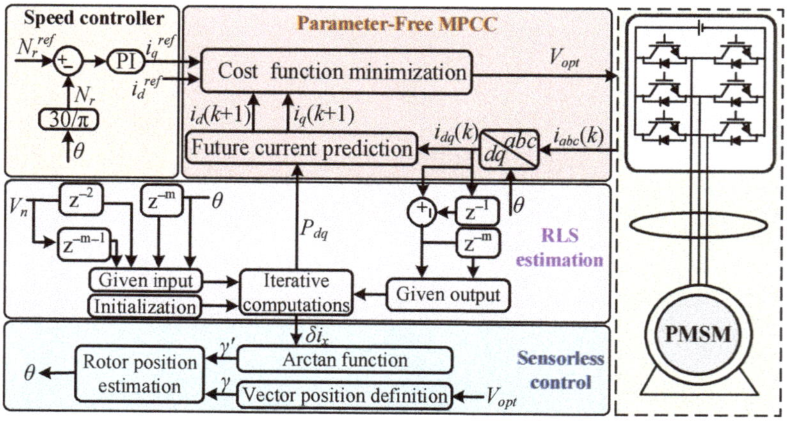

Figure 4. Structure of the proposed MPCC strategy.

1. Collection and storage of the phase currents and applied voltage vectors and obtention of $\boldsymbol{\Psi}_x(x = d,q)$ and $\boldsymbol{y}_x$ according to (15) and (16). Also, current reference $i_q^{ref}(k+1)$ needs to be achieved through the speed controller.
2. Relying on the applied voltage vector at $(k-1)$th, vector position angle γ can be decided. Meanwhile, the angle between the voltage vector and the d-axis of the rotating coordinate system, namely γ', needs to be achieved via the forced response of current variations through (26). Then, the rotor position angle θ can be estimated by (27).

3. The current variations need to be estimated through the RLS based on (23) and (24). Then, the future current can be predicted by (25).
4. The cost function g_j (10) can be evaluated and the optimal voltage vector corresponding to the minimal g_j needs to be selected to drive the inverter.

We also added the comparison results between the proposed and known strategies in Table 2.

Table 2. Comparison between the proposed and known strategies.

Reference	[14]	[17]	[18]	[26]	Proposed
Requires gain tuning	No	Yes	Yes	Yes	No
Estimation errors	Middle	Low	Low	Middle	Low
Parameter robustness	Low	Low	Low	High	High
Relative Simplicity of algorithm	Low	High	Middle	Middle	Middle

4. Experimental Results

To verify the effectiveness of proposed sensorless control strategy, this paper concerns a 1.2 kW prototype for the experiment, as presented in Figure 5. The whole hardware controller consists of the signal sampling module, the auxiliary power module, the PWM signal processing module and the minimum system board. Phase currents are measured by current sensor HAS50-S; then, current variation can be obtained. The master chip in the digital controller is TMS320F28335, and the three-phase inverter applies FF300R12ME4 IGBT modules. The sampling frequency is set as 10 kHz, and the motor parameters are all listed in Table 3.

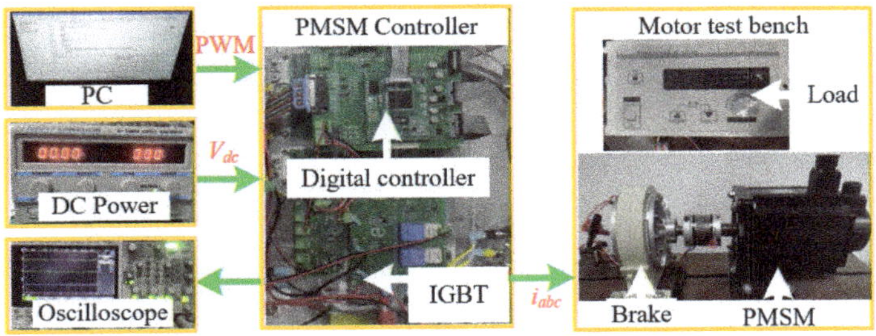

Figure 5. Hardware experiment platform.

Table 3. Motor parameters.

Parameters	Values
Rated power	1.2 kW
Rated voltage	380 V
Rated current	5 A
Rated torque	8 N·m
Rated speed	1500 rpm
Inductance of direct axis	24 mH
Inductance of quadrature axis	36 mH
Stator resistance	5.25 Ω
Moment of inertia	0.001 kg·m^2
Pole pairs	2
Permanent magnet flux linkage	0.8 Wb

In the case of parameter matching, the position estimation accuracy of the proposed control strategy is compared with that of traditional high-frequency injection position-free algorithm, in the speed commands of 50 rpm and 500 rpm, respectively. Figures 6 and 7 present the actual and estimated values, where the superscript "est" denotes estimated value. It can be seen that the estimated speed of the two sensorless control strategies is basically consistent with the actual speed, ensuring the accuracy of speed information. However, the traditional control technique has larger location estimate mistakes and more fluctuating speed because digital filters inherently have issues throughout the signal extraction process, such as signal amplitude attenuation, phase delay, and complex parameter tuning. In general, the proposed strategy has more accurate position estimation.

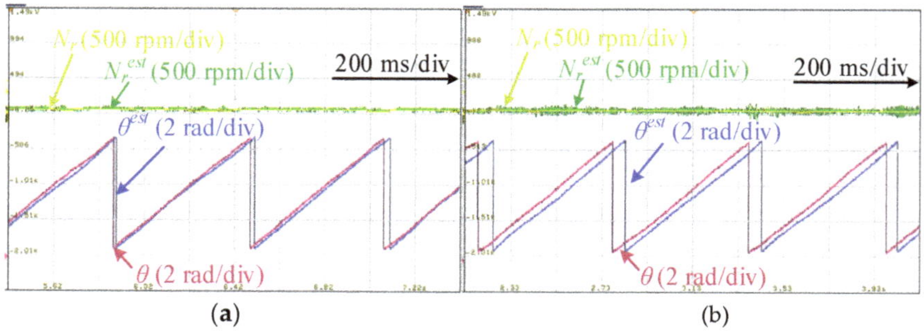

Figure 6. Comparative results between actual and estimated value at 50 rpm speed command. (**a**) Proposed control strategy. (**b**) Conventional sensorless control in [14].

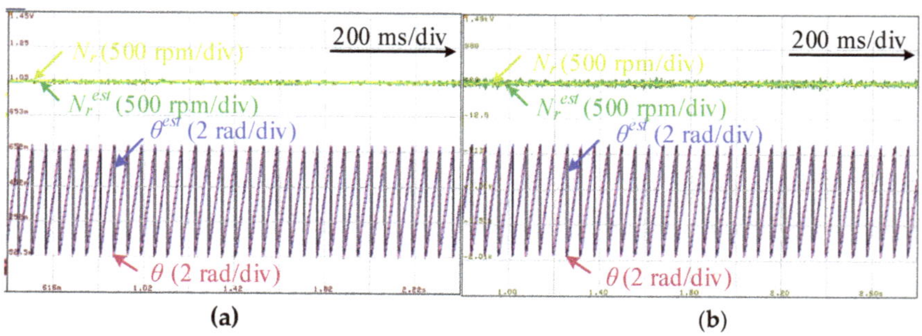

Figure 7. Comparative results between actual and estimated value at 500 rpm speed command. (**a**) Proposed control strategy. (**b**) Conventional sensorless control in [14].

To verify the robustness against parameter variations, the contrast experiments are carried out between proposed parameter-free control and model-free control in [23], when the model is mismatched and the speed command is 300 rpm. Obviously, both control strategies can maintain good sinusoidal phase current and stable rotational speed as presented in Figure 8. To be specific, the *dq*-axes prediction error of the control strategy based on RLS is basically guaranteed to be less than 0.2, while that of the control strategy based on the current variation lookup table is basically around 0.4. Thus, the future current state prediction under the proposed strategy is more accurate and the *q*-axis current ripple is smaller.

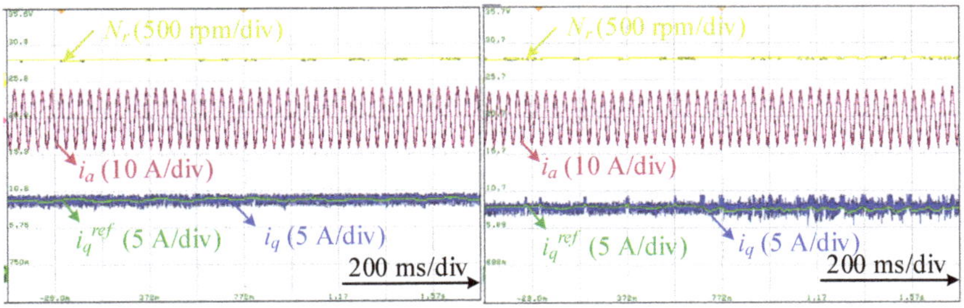

Figure 8. Experimental results of steady-state performance. (**a**) Proposed control strategy. (**b**) Parameter-free MPCC in [23].

Figure 9 further shows the contrastive steady-state results of the phase current, the torque and the *d*-axis current when the speed command is 100 rpm. Apparently, the phase currents are almost sinusoidal in two strategies. However, the model-free control in [23] fails to completely avoid the problem of stagnation of current variation update, so that it is difficult to keep the predicted current consistent with the actual current, yielding a large prediction error. Instead, the predicted current error is significantly reduced in the proposed control strategy based on the reliable estimated current variations through RLS. In general, the proposed parameter-free control strategy based on RLS reduces the influence of rotor position estimation error to some extent.

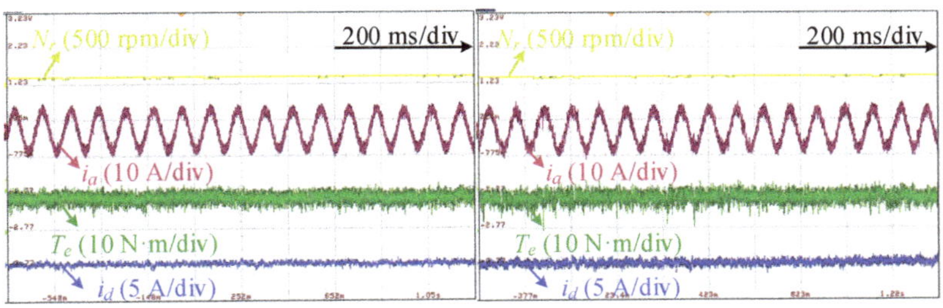

Figure 9. Experimental results of steady-state performance. (**a**) Proposed control strategy. (**b**) Parameter-free MPCC in [23].

To verify the satisfactory dynamic performance of the proposed control strategy, the experiments are carried out with variable speed and torque. Figure 10 provides the comparison results of position estimation accuracy between proposed control strategy and traditional high-frequency injection method under an accelerating condition. When the speed reference is increased from 200 rpm to 500 rpm, both control strategies exhibit similar dynamic response times. However, the position estimation error in the proposed control strategy, relying on forced response value estimation, is obviously reduced, particularly during sudden speed changes. Similarly, the comparison results of rotor position estimation are presented in Figure 11, where the load command is increased from 1 N·m to 8 N·m. It can be observed that the estimated position angle can precisely track actual values in the proposed control strategy, whereas the traditional high-frequency injection method results in noticeable discrepancies. There is no denying that the proposed control strategy performs satisfactorily under dynamic operating conditions.

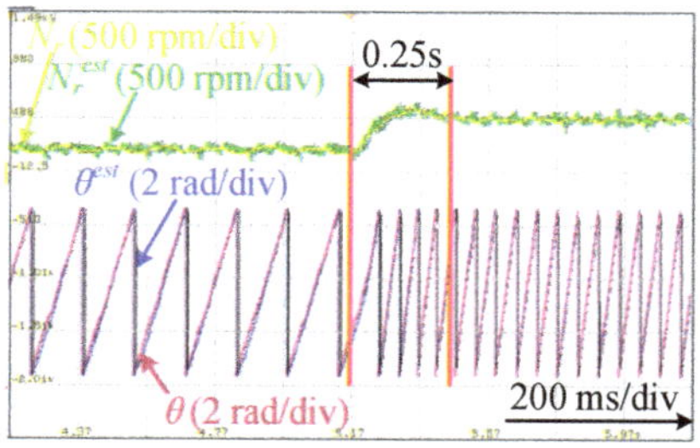

(**a**)

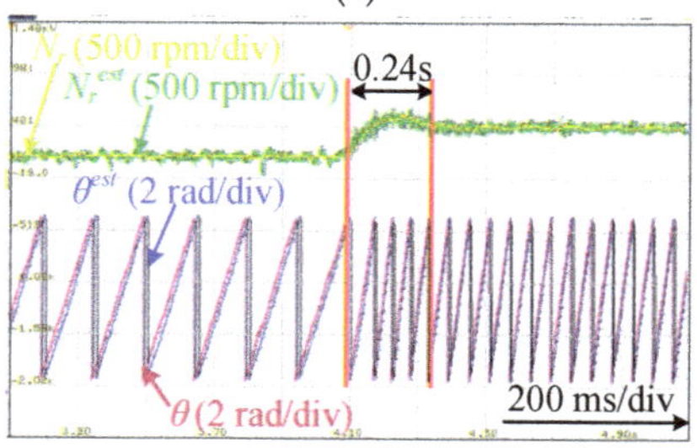

(**b**)

Figure 10. Experimental results of rotor position estimation when speed command changes. (**a**) Proposed control strategy. (**b**) Conventional sensorless control in [14].

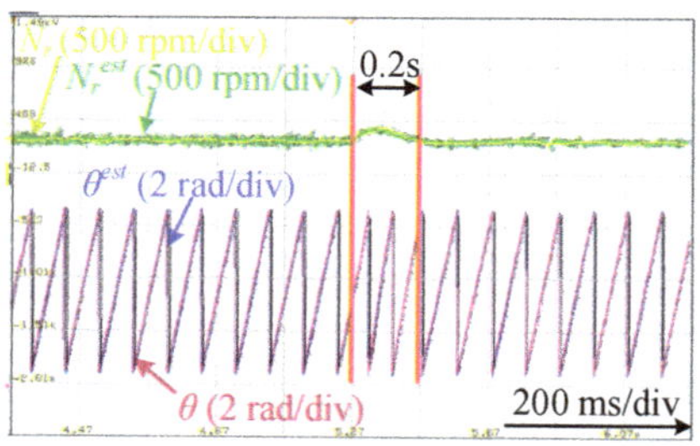

(**a**)

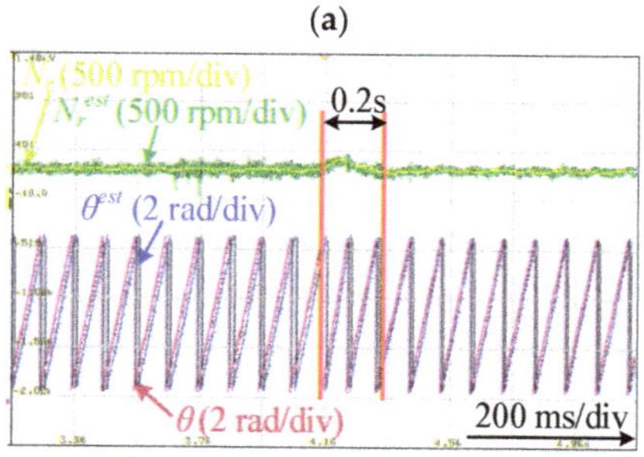

(**b**)

Figure 11. Experimental results of rotor position estimation when load command changes. (**a**) Proposed control strategy. (**b**) Conventional sensorless control in [14].

5. Conclusions

On the basis of RLS, a parameter-free MPCC control is applied for PMSM where the rotor position is obtained without any sensors. This strategy is appropriate for limited volume systems and time-varying work conditions. To be specific, the future current is no longer predicted by the mathematical model of the motor, but the estimated current variations. In addition, the relationship between the forced response and the active voltage vector in the current variation is fully utilized to obtain the rotor position. The experimental results show that the proposed control strategy can still accurately estimate the rotor position in the case of low and mutational speed command, and the estimation errors of rotor position are reduced to around 0.1 rad lower than that in [14]. When the parameters are mismatched, the proposed strategy has significantly better parameter robustness compared with that of the traditional sensorless control strategy. Simultaneously, the proposed strategy estimates the natural attenuation and the forced response by RLS in each cycle, avoiding the problem of stagnant current variation update in [23]. In general, such sensorless parameter-free MPCC broadens the application range of the PMSM. Considering the limitation of input data in RLS, in the future, we will conduct a further study to get rid of RLS algorithms so that the system can become more stable.

Author Contributions: Conceptualization, L.L. and F.Y.; methodology, L.L. and F.Y.; software, F.Y. and L.R.; validation, L.L. and F.Y.; formal analysis, L.L. and F.Y.; investigation, C.L. and F.Y.; resources, L.R. and C.L.; data curation, L.R. and C.L.; writing—original draft preparation, L.R. and C.L.; writing—review and editing, L.L. and F.Y.; visualization, L.R.; supervision, C.L. and L.R.; project administration, C.L.; All authors have read and agreed to the published version of the manuscript.

Funding: This research was funded by Natural Science Foundation of the Jiangsu Higher Education Institutions of China, grant number 22KJB470025, and Natural Science Foundation of Jiangsu Province, grant number BK20210837.

Data Availability Statement: Not applicable.

Conflicts of Interest: The authors declare no conflict of interest.

References

1. Hong, D.K.; Hwang, W.; Lee, J.Y.; Woo, B.C. Design, Analysis, and Experimental Validation of a Permanent Magnet Synchronous Motor for Articulated Robot Applications. *IEEE Trans. Magn.* **2018**, *54*, 8201304. [CrossRef]
2. Li, X.; Xue, Z.; Yan, X.; Zhang, L.; Ma, W.; Hua, W. Low-Complexity Multivector-Based Model Predictive Torque Control for PMSM with Voltage Preselection. *IEEE Trans. Power Electron.* **2021**, *36*, 11726–11738. [CrossRef]
3. Huang, W.; Hua, W.; Yin, F. Model Predictive Thrust Force Control of a Linear Flux-Switching Permanent Magnet Machine with Voltage Vectors Selection and Synthesis. *IEEE Trans. Ind. Electron.* **2019**, *66*, 4956–4967. [CrossRef]
4. Wang, W.; Liu, C.; Liu, S.; Zhao, H. Model Predictive Torque Control for Dual Three-Phase PMSMs with Simplified Deadbeat Solution and Discrete Space-Vector Modulation. *IEEE Trans. Energy Convers.* **2021**, *36*, 1491–1499. [CrossRef]
5. Zhang, X.; Li, J. Model Predictive Switching Control for PMSM Drives. *IEEE J. Emerg. Sel. Topics Power Electron.* **2023**, *11*, 942–951. [CrossRef]
6. Yu, F.; Li, K.; Zhu, Z.; Liu, X. An Over-Modulated Model Predictive Current Control for Permanent Magnet Synchronous Motors. *IEEE Access* **2022**, *10*, 40391–40401. [CrossRef]
7. Im, J.; Kim, R. Improved Saliency-Based Position Sensorless Control of Interior Permanent-Magnet Synchronous Machines with Single DC-Link Current Sensor Using Current Prediction Method. *IEEE Trans. Ind. Electron.* **2018**, *65*, 5335–5343. [CrossRef]
8. Tang, Q.; Shen, A.; Luo, P. IPMSMs Sensorless MTPA Control Based on Virtual q-Axis Inductance by Using Virtual High-Frequency Signal Injection. *IEEE Trans. Ind. Electron.* **2020**, *67*, 136–146. [CrossRef]
9. Xiao, D.; Nalakath, S.; Filho, S. Universal Full-Speed Sensorless Control Scheme for Interior Permanent Magnet Synchronous Motors. *IEEE Trans. Power Electron.* **2021**, *36*, 4723–4737. [CrossRef]
10. Teymoori, V.; Kamper, M.; Wang, R.-J.; Kennel, R. Sensorless Control of Dual Three-Phase Permanent Magnet Synchronous Machines—A Review. *Energies* **2023**, *16*, 1326. [CrossRef]
11. Fadi, A.; Ahmad, A.; Rabia, S.; Cristina, M.; Ibrahim, K. Velocity Sensor Fault-Tolerant Controller for Induction Machine using Intelligent Voting Algorithm. *Energies* **2022**, *15*, 3084.
12. Jiang, Y.; Xu, W.; Mu, C.; Zhu, J.; Dian, R. An Improved Third-Order Generalized Integral Flux Observer for Sensorless Drive of PMSMs. *IEEE Trans. Ind. Electron.* **2019**, *66*, 9149–9160. [CrossRef]
13. Chen, S.; Ding, W.; Wu, X.; Huo, L.; Hu, R.; Shi, S. Sensorless Control of IPMSM Drives Using High-Frequency Pulse Voltage Injection with Random Pulse Sequence for Audible Noise Reduction. *IEEE Trans. Power Electron.* **2023**, *38*, 9395–9408. [CrossRef]
14. Wang, G.; Yang, L.; Yuan, B.; Wang, B.; Zhang, G.; Xu, D. Pseudo-Random High-Frequency Square-Wave Voltage Injection Based Sensorless Control of IPMSM Drives for Audible Noise Reduction. *IEEE Trans. Ind. Electron.* **2016**, *63*, 7423–7433. [CrossRef]
15. Gao, S.; Yuan, L.; Dai, Y.; Mou, D.; Chen, K.; Zhao, Z. High-Frequency Current Predictive Control Method for Multiactive-Bridge Converter. *IEEE Trans. Power Electron.* **2022**, *37*, 10144–10148. [CrossRef]
16. Yang, H.; Zhang, Y.; Walker, P.D.; Liang, J.; Zhang, N.; Xia, B. Speed Sensorless Model Predictive Current Control with Ability to Start a Free Running Induction Motor. *IET Electr. Power Appl.* **2017**, *11*, 893–901. [CrossRef]

17. Kumar, P.; Bhaskar, D.V.; Muduli, U.R.; Beig, A.R.; Behera, R.K. Iron-Loss Modeling with Sensorless Predictive Control of PMBLDC Motor Drive for Electric Vehicle Application. *IEEE Trans. Transp. Electr.* **2021**, *7*, 1506–1515. [CrossRef]
18. Liu, K.; Wang, X.; Wang, R.; Sun, G.; Wang, X. Antisaturation Finite-Time Attitude Tracking Control Based Observer for a Quadrotor. *IEEE Trans. Circuits-II* **2021**, *68*, 2047–2051. [CrossRef]
19. Liu, J.; Cao, J.; Li, L. A Novel Method for Estimating the Position and Speed of a Winding Segmented Permanent Magnet Linear Motor. *Energies* **2023**, *16*, 3361. [CrossRef]
20. Wang, H.; Wu, T.; Guo, Y.; Lei, G.; Wang, X. Predictive Current Control of Sensorless Linear Permanent Magnet Synchronous Motor. *Energies* **2023**, *16*, 628. [CrossRef]
21. Eid, M.A.E.; Elbaset, A.A.; Ibrahim, H.A.; Abdelwahab, S.A.M. Modelling, Simulation of MPPT Using Perturb and Observe and Incremental Conductance techniques For Stand-Alone PV Systems. In Proceedings of the 2019 21st International Middle East Power Systems Conference (MEPCON), Cairo, Egypt, 17–19 December 2019; pp. 429–434.
22. Eid, M.A.E.; Abdelwahab, S.A.M.; Ibrahim, H.A.; Alaboudy, A.H.K. Improving the Resiliency of a PV Standalone System under Variable Solar Radiation and Load Profile. In Proceedings of the 2018 Twentieth International Middle East Power Systems Conference (MEPCON), Cairo, Egypt, 18–20 December 2018; pp. 570–576.
23. Yu, F.; Zhou, C.; Liu, X.; Zhu, C. Model-Free Predictive Current Control for Three-Level Inverter-Fed IPMSM with an Improved Current Difference Updating Technique. *IEEE Trans. Energy Convers.* **2021**, *36*, 3334–3343. [CrossRef]
24. Wu, X.; Zhu, Z.; Freire, N.M.A. High Frequency Signal Injection Sensorless Control of Finite-Control-Set Model Predictive Control with Deadbeat Solution. *IEEE Trans. Ind. Appl.* **2022**, *58*, 3685–3695. [CrossRef]
25. Wei, Y.; Xie, H.; Ke, D.; Wang, F. Sensorless Model-Free Predictive Current Control with Variable Prediction Horizon by Estimated Position for PMSM. In Proceedings of the 2022 25th International Conference on Electrical Machines and Systems (ICEMS), Chiang Mai, Thailand, 29 November–2 December 2022; pp. 1–5.
26. Zhu, X.; Zhang, L.; Xiao, X.; Lee, C.H.T.; Que, H. Adjustable-Flux Permanent Magnet Synchronous Motor Sensorless Drive System Based on Parameter-Sensitive Adaptive Online Decoupling Control Strategy. *IEEE Trans. Transp. Electr.* **2023**, *9*, 501–511. [CrossRef]

energies

Article

High-Resistance Connection Diagnosis of Doubly Fed Induction Generators

Wei Ding [1,2], Yulong Jin [1,2], Xijin Wu [1,2], Yufeng Yang [1,2,*] and Yongjiang Jiang [3]

[1] NARI Group Corporation, State Grid Electric Power Research Institute, Nanjing 211106, China; dingwei1@sgepri.sgcc.com.cn (W.D.); jinyulong@sgepri.sgcc.com.cn (Y.J.); wuxijin@sgepri.sgcc.com.cn (X.W.)
[2] NARI Technology Co., Ltd., Nanjing 211106, China
[3] College of Automation & College of Artificial Intelligence, Nanjing University of Posts and Telecommunications, Nanjing 210023, China; yongj@njupt.edu.cn
* Correspondence: yangyufeng@sgepri.sgcc.com.cn

Abstract: The high resistance connection fault of the stator is a common fault in doubly fed induction generators, which causes a three-phase imbalance in the stator circuit. Since the stator winding is directly connected to the power grid, interference from the asymmetric power grid must be eliminated in order to achieve the accurate diagnosis of stator resistance imbalance faults. Therefore, a new diagnosis method based on filter shunt capacitor banks is proposed in this paper. By introducing shunt capacitor banks, an artificial neutral point is constructed to replace the neutral point of the power grid. Then, the neutral point voltage of the stator winding relative to the artificial neutral point is selected as a fault characteristic signal. In this paper, the change in three-phase stator currents after a high-resistance connection fault is analyzed in detail, and by comparing the fault characteristic signal with three-phase stator currents, the fault phase location and fault severity of high-resistance connection can be accurately obtained. Finally, simulations are carried out via the field-circuit coupling method to validate the effectiveness of the proposed method.

Keywords: doubly fed induction generator; fault diagnosis; high-resistance connection; neutral point voltage; shunt capacitor banks

Citation: Ding, W.; Jin, Y.; Wu, X.; Yang, Y.; Jiang, Y. High-Resistance Connection Diagnosis of Doubly Fed Induction Generators. *Energies* **2023**, *16*, 7516. https://doi.org/10.3390/en16227516

Academic Editor: Javier Contreras

Received: 12 September 2023
Revised: 1 November 2023
Accepted: 7 November 2023
Published: 9 November 2023

1. Introduction

1.1. Motivations

At present, due to the use of partial capacity power converters, the doubly fed induction generator (DFIG) is still one of the most widely used wind-driven generators. With the rapid development of wind power, its installed capacity continues to increase. However, wind turbines typically operate in harsh environments, which may cause unexpected faults in the generator. High-resistance connection (HRC) is one of the common electrical faults of the generator [1,2]. Figure 1 shows the HRC fault mechanism of DFIG [3]. HRC faults lead to a decrease in the performance of the generator, such as imbalanced electrical signals in the stator windings, overheating, increased torque pulsation, and a reduction in the effective electromagnetic torque [4,5]. Severe cases evolve into open circuit faults and even lead to damage to the entire wind turbine. Therefore, it is necessary to detect and repair HRC faults as early as possible to avoid further evidence of the fault.

1.2. Related Works

Currently, many achievements have been made in the detection of HRC faults in induction motors internationally [6–9]. Based on analyzing and processing the external magnetic field during motor startup, ref. [6] characteristic signals can be obtained indicating HRC faults in the rotor winding. However, this method needs to be carried out during startup, so it can only achieve offline detection. The study in [7] calculates the maximum energy density of high-order fault harmonic signals under the startup transient as training samples and uses feedforward neural networks for rotor fault degree classification. This method is only suitable for judging the degree of a fault during startup and requires strong signal processing and computational capabilities. Study [8] proposes a method that can diagnose stator winding HRC faults in wound induction motors by comprehensively analyzing the steady-state and transient currents of the rotor winding. This method not only requires the collection of stable operating currents but also the collection of transient starting currents. The study in [9] calculates the bi-coherence of the stray flux signal during startup for fuzzy c-means machine learning to realize the rotor winding fault diagnosis.

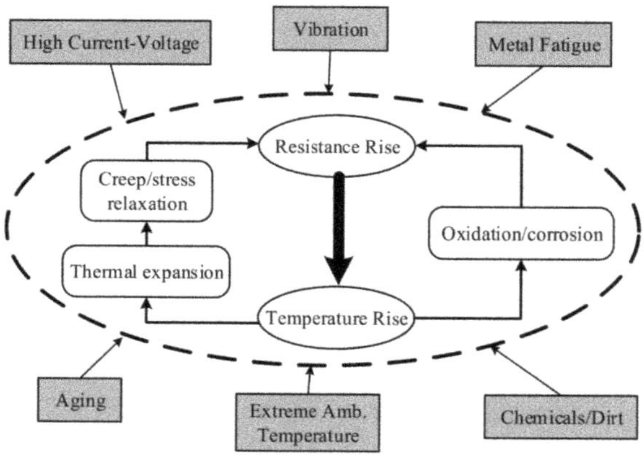

Figure 1. HRC fault mechanism of DFIG.

In recent years, the literature on the HRC fault detection of permanent magnet synchronous motors (PMSM) has been constantly emerging. In [10–12], zero sequence voltage is obtained by attaching a resistor network to detect the HRC faults; meanwhile, the fault phase and degree can be determined. In [13], to realize the HRC fault diagnosis of the PMSM with the direct torque control, two pairs of bias magnetic links with different pole numbers are superimposed, and the resistance deviation is obtained by solving a binary linear equation system. In [14], voltage distortion is estimated to detect the HRC fault using a reference model. Similarly, a high-order sliding mode controller is proposed in [15] to achieve HRC fault–tolerant control and fault severity estimation. In [16,17], two deep learning algorithms, Deep Neural Networks, and deep Q-network, are used for the intelligent diagnosis of winding faults in PMSMs.

Unlike induction motors and PMSM, DFIG has dual electrical ports, where the rotor winding is connected to the grid through a power converter, while stator winding is directly connected to the grid. The electrical environment of the stator winding is completely different from other machines, and existing methods are not applicable. Meanwhile, since the stator winding is directly connected to the power grid; the imbalance of the grid disturbs the diagnosis. Studies [18–21] proposed some methods to detect the rotor HRC fault in DFIGs; however, there is limited research on the HRC diagnosis of the stator winding.

1.3. Contributions

Based on the relevant literature, the HRC fault diagnosis methods can be roughly classified into the following three categories: model analysis-based methods [4,14,15,21], signal processing-based methods [5,6,8,10–13,18,20], and knowledge-based methods [7,9,16,17,19]. Among them, the model analysis method delves into the fault model and mechanism of machines, but, in practical application, it relies on accurate mathematical models. However, the parameters of wind turbines may change at different operating points, which can easily lead to significant errors in parameter or state estimation results, leading to misdiagnosis and low reliability. The signal processing-based method is currently the most widely used in the field of fault diagnosis for wind turbines, which, to some extent, avoids the problem of mathematical models of diagnostic objects. However, it also needs to address noise and complexity issues to improve the efficiency of signal processing. Knowledge-based methods have emerged with the rapid development of computer technology and artificial intelligence technology. It does not require the establishment of an accurate mathematical model for the diagnostic object, but only provides various data on the motor's operating status. However, the realization of knowledge-based methods needs the acquisition of prior knowledge and data, and the diagnostic performance directly depends on the amount of fault sample data. In addition, the significant impact of the operating status on motor fault characteristics, to some extent, increases the difficulty of intelligent motor fault diagnosis.

In this paper, based on the shunt capacitor banks and established artificial neutral points, a new HRC diagnosis method for the stator winding of DFIG is proposed. The proposed method is a combination of a model analysis-based method and data processing-based method. The proposal of this method is based on the analysis of the DFIG fault model, but the analysis of the model is only to seek a fault characteristic quantity that is not affected by the model accuracy, generator parameters, and generator operating conditions. Meanwhile, only one fundamental signal needs to be extracted in the proposed method as a data processing method is simple and efficient. And both single-phase and multiphase faults can be accurately located and evaluated simultaneously. This means it has great practical application potential.

1.4. Paper Organization

The organizational structure of this paper is arranged as follows: In Section 2, the model of DIFG with HRC is established. In Section 3, the potential drift of the stator winding neutral point using HRC is analyzed, and the HRC diagnosis method based on the artificial neutral point is proposed. In Section 4, simulations are conducted to verify the proposed diagnostic method. In Section 5, the simulation results were discussed. Finally, conclusions and the conclusion and outlook for future work are drawn in Section 6.

2. DFIG Model with HRC

The HRC in the stator winding of DFIG can be simulated by connecting an additional resistor in series to the faulty phase of the stator winding. It is assumed that the HRC occurs in phase A, as shown in Figure 2.

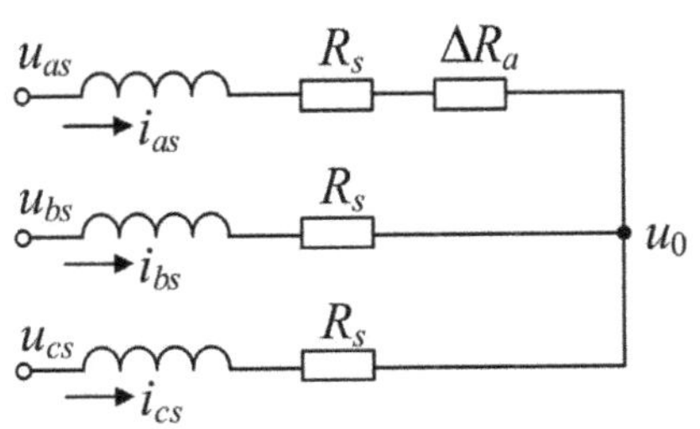

Figure 2. Equivalent model of DFIG stator winding with the HRC in phase A.

In Figure 2, u_{as}, u_{bs}, and u_{cs} are the terminal voltages of stator three-phase winding, respectively; u_0 is the neutral point voltage of the stator winding; i_{as}, i_{bs}, and i_{cs} are the currents of the stator three-phase winding, respectively; R_s is the phase resistance of the stator winding; ΔR_a is the additional resistance in phase A due to the HRC. According to the operating conditions in Figure 2, ignoring the harmonics, the voltage and flux equations of the DFIG with HRC in stator phase A are derived as follows:

$$\begin{bmatrix} u_s \\ u_r \end{bmatrix} = \begin{bmatrix} R_s & 0 \\ 0 & R_r \end{bmatrix} \begin{bmatrix} i_s \\ i_r \end{bmatrix} + \frac{d}{dt}\left(\begin{bmatrix} \psi_s \\ \psi_r \end{bmatrix} \right) + \begin{bmatrix} u_{s0} \\ 0 \end{bmatrix} \tag{1}$$

$$\begin{bmatrix} \psi_s \\ \psi_r \end{bmatrix} = \begin{bmatrix} L_{ss} & M_{sr} \\ M_{sr}^{\mathrm{T}} & L_{rr} \end{bmatrix} \begin{bmatrix} i_s \\ i_r \end{bmatrix} \tag{2}$$

where $[u_s]$ and $[u_r]$ are the voltage matrices of the stator winding and rotor winding, respectively; $[u_{s0}] = [u_0\ u_0\ u_0]^{\mathrm{T}}$; $[i_s]$ and $[i_r]$ are the current matrices of the stator winding and rotor winding, respectively; $[R_s]$ and $[R_r]$ are the resistance matrices of the stator winding and rotor winding, respectively; $[\psi_s]$ and $[\psi_r]$ are the flux matrices of the stator winding and rotor winding, respectively; $[L_{ss}]$ and $[L_{rr}]$ are the inductance matrices of the stator winding and rotor winding, respectively; and $[M_{sr}]$ is the mutual inductance matrix of the stator and rotor windings. Additionally, their expressions are as follows:

$$[u_s] = [u_{as} \quad u_{bs} \quad u_{cs}]^{\mathrm{T}} \tag{3}$$

$$[u_r] = [u_{ar} \quad u_{br} \quad u_{cr}]^{\mathrm{T}} \tag{4}$$

$$[i_s] = [i_{as} \quad i_{bs} \quad i_{cs}]^{\mathrm{T}} \tag{5}$$

$$[i_r] = [i_{ar} \quad i_{br} \quad i_{cr}]^{\mathrm{T}} \tag{6}$$

$$[R_s] = \begin{bmatrix} R_s + \Delta R_a & 0 & 0 \\ 0 & R_s & 0 \\ 0 & 0 & R_s \end{bmatrix} \tag{7}$$

$$[R_r] = \begin{bmatrix} R_r & 0 & 0 \\ 0 & R_r & 0 \\ 0 & 0 & R_r \end{bmatrix} \tag{8}$$

$$[L_{ss}] = \begin{bmatrix} L_s & M_s & M_s \\ M_s & L_s & M_s \\ M_s & M_s & L_s \end{bmatrix} \tag{9}$$

$$[L_{rr}] = \begin{bmatrix} L_r & M_r & M_r \\ M_r & L_r & M_r \\ M_r & M_r & L_r \end{bmatrix} \tag{10}$$

$$[M_{sr}] = L_{sr} \begin{bmatrix} \cos\theta_r & \cos(\theta_r + \frac{2}{3}\pi) & \cos(\theta_r - \frac{2}{3}\pi) \\ \cos(\theta_r - \frac{2}{3}\pi) & \cos\theta_r & \cos(\theta_r + \frac{2}{3}\pi) \\ \cos(\theta_r + \frac{2}{3}\pi) & \cos(\theta_r - \frac{2}{3}\pi) & \cos\theta_r \end{bmatrix} \tag{11}$$

where u_{ar}, u_{br}, and u_{cr} are the terminal voltages of the rotor three-phase winding, respectively; i_{ar}, i_{br}, and i_{cr} are the currents of the rotor three-phase winding, respectively; R_r is the phase resistance of the rotor winding; L_s and M_s are the self-inductance and mutual inductance of the stator winding, respectively, and $M_s = -1/2\ L_s$; L_r and M_r are the self-inductance and mutual inductance of the rotor winding, respectively, and $M_r = -1/2\ L_r$; L_{sr} is the maximum mutual inductance of the stator and rotor windings; and θ_r is the electrical angle difference of the A-phase winding axis of the stator and rotor, which can be obtained via the following:

$$\theta_r = p\omega_r t + \theta_0 \tag{12}$$

where θ_0 is the initial electrical angle difference of the A-phase winding axis of the stator and rotor; ω_r is the electrical angular velocity of the rotor; and p is the pole-pair number of the DFIG.

3. Diagnostic Method

3.1. Potential Drift of Stator Winding Neutral Point

From the above mathematical model, it can be seen that the HRC of the stator winding did not change the voltage equation of the rotor winding. According to (1) and (7), the HRC leads to three-phase asymmetry in the stator winding, which is directly connected to the power grid, causing the potential drift of the stator winding neutral point. Then, the stator voltage equation can be expressed as follows:

$$\begin{bmatrix} u_{as} \\ u_{bs} \\ u_{cs} \end{bmatrix} = \begin{bmatrix} R_s + \Delta R_a & 0 & 0 \\ 0 & R_s & 0 \\ 0 & 0 & R_s \end{bmatrix} \begin{bmatrix} i_{as} \\ i_{bs} \\ i_{cs} \end{bmatrix} + \frac{d}{dt}\left(\begin{bmatrix} L_s & M_s & M_s \\ M_s & L_s & M_s \\ M_s & M_s & L_s \end{bmatrix} \begin{bmatrix} i_{as} \\ i_{bs} \\ i_{cs} \end{bmatrix} \right) + \begin{bmatrix} e_{as} \\ e_{bs} \\ e_{cs} \end{bmatrix} + \begin{bmatrix} u_0 \\ u_0 \\ u_0 \end{bmatrix} \tag{13}$$

where e_{as}, e_{bs}, and e_{cs} are the stator back-EMFs of the DFIG, which are induced by the three-phase currents of the rotor. It produces the following:

$$\begin{bmatrix} e_{as} \\ e_{bs} \\ e_{cs} \end{bmatrix} = \frac{d}{dt}\left(L_{sr} \begin{bmatrix} \cos\theta_r & \cos(\theta_r + \frac{2}{3}\pi) & \cos(\theta_r - \frac{2}{3}\pi) \\ \cos(\theta_r - \frac{2}{3}\pi) & \cos\theta_r & \cos(\theta_r + \frac{2}{3}\pi) \\ \cos(\theta_r + \frac{2}{3}\pi) & \cos(\theta_r - \frac{2}{3}\pi) & \cos\theta_r \end{bmatrix} \begin{bmatrix} i_{ar} \\ i_{br} \\ i_{cr} \end{bmatrix} \right) \tag{14}$$

Since the rotor winding of the DFIG is driven by the inverter, the rotor currents can be controlled as three-phase symmetrical sinusoidal currents, which can be expressed as follows:

$$\begin{bmatrix} i_{ar} \\ i_{br} \\ i_{cr} \end{bmatrix} = I_r \begin{bmatrix} \cos\omega_c t \\ \cos(\omega_c t - \frac{2}{3}\pi) \\ \cos(\omega_c t + \frac{2}{3}\pi) \end{bmatrix} \tag{15}$$

where I_r is the amplitude of rotor current; ω_c is the angular frequency of rotor currents. According to the operating principle of the DFIG, it can express the following:

$$\omega_s = p\omega_r + \omega_c \tag{16}$$

where ω_s is the angular frequency of stator currents. Substituting (12), (15), and (16) into (14) obtains the following:

$$\begin{bmatrix} e_{as} \\ e_{bs} \\ e_{cs} \end{bmatrix} = \frac{3}{2} L_{sr} I_r \begin{bmatrix} \cos(\omega_s t + \theta_0) \\ \cos(\omega_s t + \theta_0 - \frac{2}{3}\pi) \\ \cos(\omega_s t + \theta_0 + \frac{2}{3}\pi) \end{bmatrix} \tag{17}$$

Based on Kirchhoff's current law, it produces the following:

$$i_{as} + i_{bs} + i_{cs} = 0 \tag{18}$$

According to (13), (17), and (18), the potential drift of the stator winding neutral point can be expressed as

$$u_0 = \frac{1}{3}(u_{as} + u_{bs} + u_{cs}) - \frac{1}{3}\Delta R_a i_{as} \tag{19}$$

3.2. HRC Diagnosis under Balanced Grid

Since the stator winding of the DFIG is connected to the power grid directly, u_{as}, u_{bs}, and u_{cs} are also the three-phase voltages of the power grid, as shown in Figure 3. In a three-phase balanced grid, it can be expressed as:

$$u_{as} + u_{bs} + u_{cs} = 0 \tag{20}$$

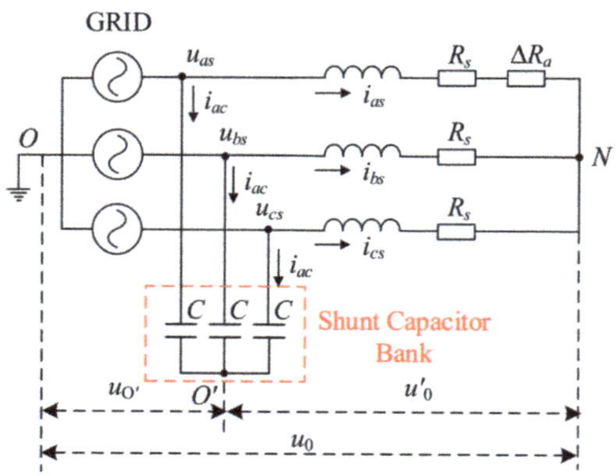

Figure 3. Connection diagram of DFIG stator winding.

When substituting (20) into (19), we can obtain

$$\begin{aligned} u_0 &= -\tfrac{1}{3}\Delta R_a i_{as} = -\tfrac{1}{3}\Delta R_a I_{as}\cos(\omega_s t + \theta_a) \\ &= \tfrac{1}{3}\Delta R_a I_{as}\cos(\omega_s t + \theta_a + \pi) \\ &= U_0 \cos(\omega_s t + \alpha) \end{aligned} \tag{21}$$

where I_{as} and θ_a are the current amplitude and initial phase of stator A-phase winding, respectively. U_0 and α are the voltage amplitude and initial phase of the stator winding neutral point, and

$$\begin{cases} U_0 = \tfrac{1}{3}\Delta R_a I_{as} \\ \alpha = \theta_a + \pi \end{cases} \tag{22}$$

From (22), it can be seen that the amplitude and initial phase angle of the neutral point potential u_0 are directly related to the amplitude and initial phase angle of the stator fault phase current i_{as} as well as additional resistance ΔR_a.

Similarly, if the HRC occurs in phase B or phase C, it can be obtained as follows:

$$\begin{cases} U_0 = \tfrac{1}{3}\Delta R_b I_{bs} \\ \alpha = \theta_b + \pi \end{cases} \text{or} \begin{cases} U_0 = \tfrac{1}{3}\Delta R_c I_{cs} \\ \alpha = \theta_c + \pi \end{cases} \tag{23}$$

More generally, when HRC occurs simultaneously in two-phase windings, such as phase A and phase B, it produces the following:

$$\begin{aligned} u_0 &= -\Delta R_a i_{as} - \Delta R_b i_{bs} \\ &= -\tfrac{1}{3}\Delta R_a I_{as}\cos(\omega_s t + \theta_a) - \tfrac{1}{3}\Delta R_b I_{bs}\cos(\omega_s t + \theta_b) \\ &= A\cos(\omega_s t + \theta_a + \pi) + B\cos(\omega_s t + \theta_b + \pi) \\ &= U_0\cos(\omega_s t + \alpha) \end{aligned} \tag{24}$$

where

$$\begin{cases} U_0 = \sqrt{A^2 + B^2 + 2AB\ \cos(\theta_a - \theta_b)} \\ A = \tfrac{1}{3}\Delta R_a I_{as} \\ B = \tfrac{1}{3}\Delta R_b I_{bs} \\ \alpha = \arccos\left(\frac{A\ \cos(\theta_a+\pi)+B\ \cos(\theta_b+\pi)}{U_0}\right) \end{cases} \tag{25}$$

3.3. Artificial Neutral Point

The above analysis is based on the three-phase balance of the power grid. When the three-phase imbalance occurs in the power grid, (20) is valid, and then (21)~(25) are no longer applicable. In this case, only (19) could be applied to calculate the neutral point potential u_0. As can be seen from (19), u_0 includes the following two parts: a three-phase imbalance in the power grid and a three-phase imbalance in the stator winding. And then, it is no longer possible to use u_0 to determine whether or where HRC occurs. In order to eliminate the impact of the power grid imbalance, an artificial neutral point is constructed to replace the neutral point of the power grid in this paper.

Considering that, in wind power generation systems, three-phase symmetrical capacitor banks are usually connected in parallel to achieve filtering and reactive power compensation functions, their neutral point is easily obtained. Then, the neutral point of the shunt capacitor bank O' is introduced to replace the neutral point of the power grid O, as shown in Figure 3. The voltage between the two neutral points of the power grid and capacitor bank is $u_{o'}$. Then, the current equation of the shunt capacitor bank can be expressed as follows:

$$\begin{bmatrix} i_{ac} \\ i_{bc} \\ i_{cc} \end{bmatrix} = C\frac{d}{dt}\left(\begin{bmatrix} u_{as} \\ u_{bs} \\ u_{cs} \end{bmatrix} - \begin{bmatrix} u'_{0} \\ u'_{0} \\ u'_{0} \end{bmatrix}\right) \quad (26)$$

where C is the capacitance value in the parallel capacitor bank.

Based on Kirchhoff's current law, it can be expressed as

$$i_{ac} + i_{bc} + i_{cc} = 0 \quad (27)$$

Substituting (27) into (26), the following is produced:

$$u_{o'} = \frac{1}{3}(u_{as} + u_{bs} + u_{cs}) + K \quad (28)$$

where K is a constant. In order to ensure that (26) remains true, K must be an invariant constant. And since the initial value of K is 0, it must be 0. Then, the voltage $u_{o'}$ is as follows:

$$u_{o'} = \frac{1}{3}(u_{as} + u_{bs} + u_{cs}) \quad (29)$$

Then, according to (29), the neutral point of the parallel capacitor bank can also be used to detect three-phase unbalanced faults in the grid.

According to (19) and (29), it is expressed as

$$u'_{0} = u_0 - u_{o'} = -\frac{1}{3}\Delta R_a i_{as} \quad (30)$$

where u'_0 is the voltage between the two neutral points of the stator winding and shunt capacitor bank. If u'_0 is used instead of u_0 in (21) and (24), (21)~(25) are always correct regardless of whether the three phases of the power grid are balanced or not.

3.4. Fault Location and Degree Estimation

According to the above analysis, the HRC faults can be identified by detecting u'_0, which is expressed as follows:

$$\begin{cases} u'_0 = U_0\cos(\omega_s t + \alpha) \\ U_0 = \sqrt{X^2 + Y^2 + 2XY\ \cos(\theta_x - \theta_y)} \\ X = \frac{1}{3}\Delta R_x I_{xs} \\ Y = \frac{1}{3}\Delta R_y I_{ys} \\ \alpha = \arccos\left(\frac{X\ \cos(\theta_x + \pi) + Y\ \cos(\theta_y + \pi)}{U_0}\right) \end{cases} \quad (31)$$

where $x = a, y = b$; or $x = b, y = c$; or $x = c, y = a$.

Then, when the HRC fault occurs in one phase winding of the stator, u'_0 can be expressed as follows:

$$\begin{cases} u'_0 = U_0\cos(\omega_s t + \alpha) \\ U_0 = X = \frac{1}{3}\Delta R_x I_{xs} \\ \alpha = \theta_x + \pi \end{cases} \tag{32}$$

where $x = a, b$, or c. It can be seen that an HRC fault in one-phase stator winding can be considered as a special case of the HRC fault in two-phase stator windings. The fault degree and the faulty phase can be conveniently estimated using U_0 and α, respectively. If the HRC occurs in all three phase windings of the stator, since the HRC causes an asymmetric fault, the $\min(\Delta R_a, \Delta R_b, \Delta R_c)$ is considered as part of the normal resistance value of the stator windings, and U_0 and α can be calculated as follows:

$$\begin{cases} U_0 = \sqrt{X^2 + Y^2 + 2XY\ \cos(\theta_x - \theta_y)} \\ X = \frac{1}{3}(\Delta R_x - \Delta R_{\min})I_{xs} \\ Y = \frac{1}{3}(\Delta R_y - \Delta R_{\min})I_{ys} \\ \Delta R_{\min} = \min(\Delta R_a, \Delta R_b, \Delta R_c) \\ \alpha = \arccos(\frac{X\ \cos(\theta_x+\pi)+Y\ \cos(\theta_y+\pi)}{U_0}) \end{cases} \tag{33}$$

It can be seen that HRC faults in the three-phase stator winding can be transformed into HRC faults in the two-phase stator winding. Therefore, for the convenience of analysis, whether the fault occurs in a single-phase, two-phase, or three-phase stator winding, it is always assumed that the HRC fault occurs in the two-phase stator winding. Meanwhile, U_0 and α are selected as fault characteristics, which can be obtained by gathering the voltage signal between the neutral point N and O'. According to (31)~(33), since the fault phase information cannot be obtained only through α, it is necessary to compare it with the phase of the three-phase stator current and determine the fault location through the phase relationship between u'_0 and the three-phase stator current. Therefore, $\theta_{\alpha a}$, $\theta_{\alpha b}$, and $\theta_{\alpha c}$ are used to replace α as fault characteristics and can directly locate faults based on their values. They can be obtained using the following:

$$\begin{cases} \theta_{\alpha x} = |\alpha - \theta_x - \pi| & |\alpha - \theta_x - \pi| \le \pi \\ \theta_{\alpha x} = 2\pi - |\alpha - \theta_x - \pi| & |\alpha - \theta_x - \pi| > \pi \end{cases} \tag{34}$$

where $x = a, b$, or c. Then, by collecting the amplitude and phase of the three-phase stator current, combining (31)~(34), the fault location and degree can be calculated.

In order to achieve rapid fault localization, the range of feature values for different fault phases needs to be analyzed. For the sake of analysis, it is still assumed that phase A is the faulty phase, then according to (32) and (34), $\theta_{\alpha a} = 0$, $\theta_{\alpha b} = \theta_a - \theta_b$, and $\theta_{\alpha c} = \theta_c - \theta_a$. In the healthy generator system, the three-phase currents are symmetrical, and the phase difference between them is $2\pi/3$. However, in the case of HRC, the three-phase currents are no longer symmetrical, and the phase relationship of $2\pi/3$ between them is no longer satisfied. Figure 4 shows the phasor diagram of the DFIG stator winding with HRC in phase A, where the red font and line segments represent the phasor that changes after the HRC occurs, while the black represents the phasor before the HRC occurs. In Figure 4, similar to the non-salient pole synchronous generator, the synchronous reactance X_s is defined, and expressed as follows:

$$X_s = \frac{3}{2}\omega_s L_s \tag{35}$$

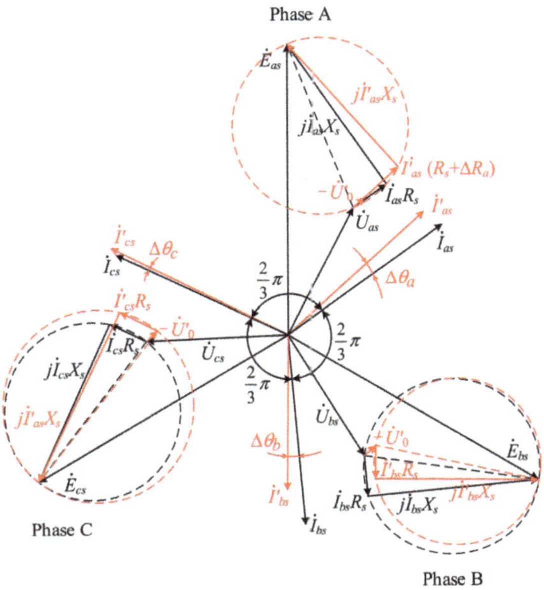

Figure 4. Phasor diagram of DFIG stator winding with HRC in phase A.

As shown in Figure 4, due to grid clamping, the voltage $\dot{U}_{as}$, $\dot{U}_{bs}$ and $\dot{U}_{cs}$ does not change after the HRC fault. According to (17), the EMFs $\dot{E}_{as}$, $\dot{E}_{bs}$ and $\dot{E}_{cs}$ do not change using the HRC. Then, three phase voltages $\dot{E}_{as}-\dot{U}_{as}$, $\dot{E}_{bs}-\dot{U}_{bs}$ and $\dot{E}_{cs}-\dot{U}_{cs}$ are still symmetrical after HRC occurs. With them as a reference, it is convenient to calculate the change in the phase and amplitude of three-phase currents after the HRC fault. According to Figure 4, it can be expressed as follows:

$$\begin{cases} \Delta\theta_a = \arctan\frac{X_s}{R_s} - \arctan\frac{X_s}{R_s+\frac{4}{3}\Delta R_a} \\ I'_{as} = \frac{\sqrt{R_s^2+X_s^2}}{\sqrt{(R_s+\frac{4}{3}\Delta R_a)^2+X_s^2}} I_m \\ \Delta\theta_b = \frac{2\pi}{3} - \arcsin\frac{3\sqrt{3}R_s+4\sqrt{3}\Delta R_a+3X_s}{\sqrt{(6R_s+9\Delta R_a)^2+(6X-\sqrt{3}\Delta R_a)^2}} - \arctan\frac{X_s}{R_s+\frac{4}{3}\Delta R_a} & \Delta R_a \le \frac{\sqrt{3}X_s-R_s}{2} \\ \Delta\theta_b = \arcsin\frac{3\sqrt{3}R_s+4\sqrt{3}\Delta R_a+3X_s}{\sqrt{(6R_s+9\Delta R_a)^2+(6X-\sqrt{3}\Delta R_a)^2}} - \arctan\frac{X_s}{R_s+\frac{4}{3}\Delta R_a} - \frac{\pi}{3} & \Delta R_a > \frac{\sqrt{3}X_s-R_s}{2} \\ I'_{bs} = \frac{\sqrt{(6R_s+9\Delta R_a)^2+(6X-\sqrt{3}\Delta R_a)^2}}{2\sqrt{(3R_s+4\Delta R_a)^2+9X_s^2}} I_m \end{cases} \quad (36a)$$

$$\begin{cases} \Delta\theta_c = \frac{\pi}{3} + \arcsin\frac{-3\sqrt{3}R_s-4\sqrt{3}\Delta R_a+3X_s}{\sqrt{(6R_s+9\Delta R_a)^2+(6X+\sqrt{3}\Delta R_a)^2}} - \arctan\frac{X_s}{R_s+\frac{4}{3}\Delta R_a} & \Delta R_a \le \frac{\sqrt{3}X_s-3R_s}{4} \\ \Delta\theta_c = \frac{\pi}{3} - \arcsin\frac{-3\sqrt{3}R_s-4\sqrt{3}\Delta R_a+3X_s}{\sqrt{(6R_s+9\Delta R_a)^2+(6X+\sqrt{3}\Delta R_a)^2}} - \arctan\frac{X_s}{R_s+\frac{4}{3}\Delta R_a} & \Delta R_a > \frac{\sqrt{3}X_s-3R_s}{4} \\ I'_{cs} = \frac{\sqrt{(6R_s+9\Delta R_a)^2+(6X+\sqrt{3}\Delta R_a)^2}}{2\sqrt{(3R_s+4\Delta R_a)^2+9X_s^2}} I_m \end{cases} \quad (36b)$$

where $\Delta\theta_a$, $\Delta\theta_b$, and $\Delta\theta_c$ are the phase offset of the three-phase stator currents when the HRC fault occurs in phase A. I'_{as}, I'_{bs}, and I'_{cs} are the amplitude of the three-phase stator currents of the DFIG with HRC in phase A. I_m is the amplitude of the stator currents of the healthy DFIG.

Clearly, with the deepening of the HRC fault, the phase difference between the three-phase fault current and healthy current constantly changes. When HRC occurs in phase A, as the HRC fault deepens, according to (36), $\Delta\theta_a$ changes from 0 to $\arctan X_s/R_s$, $\Delta\theta_b$ changes from 0 to $-11\pi/180$, and $\Delta\theta_c$ changes from 0 to $11\pi/180$. Meanwhile, $\Delta\theta_c$ does not monotonically increase with the increase in the fault severity. As shown in Figure 4, when the degree of fault is very small, $\Delta\theta_c$ might be slightly less than 0, and the specific value is determined by the parameters of the DFIG.

According to (36), it can be seen that both the amplitude and phase of the three-phase currents are no longer symmetrical when HRC occurs. The phase fault characteristics can be calculated as follows:

$$\begin{cases} \theta_{\alpha a} = 0 & \\ \theta_{\alpha b} = \theta_a - \theta_b = \arcsin\frac{3\sqrt{3}R_s+4\sqrt{3}\Delta R_a+3X_s}{\sqrt{(6R_s+9\Delta R_a)^2+(6X-\sqrt{3}\Delta R_a)^2}} + \arctan\frac{X_s}{R_s} & \Delta R_a \le \frac{\sqrt{3}X_s-R_s}{2} \\ \theta_{\alpha b} = \theta_a - \theta_b = \pi - \arcsin\frac{3\sqrt{3}R_s+4\sqrt{3}\Delta R_a+3X_s}{\sqrt{(6R_s+9\Delta R_a)^2+(6X-\sqrt{3}\Delta R_a)^2}} + \arctan\frac{X_s}{R_s} & \Delta R_a > \frac{\sqrt{3}X_s-R_s}{2} \\ \theta_{\alpha c} = \theta_c - \theta_a = \pi + \arcsin\frac{-3\sqrt{3}R_s-4\sqrt{3}\Delta R_a+3X_s}{\sqrt{(6R_s+9\Delta R_a)^2+(6X+\sqrt{3}\Delta R_a)^2}} - \arctan\frac{X_s}{R_s} & \Delta R_a \le \frac{\sqrt{3}X_s-3R_s}{4} \\ \theta_{\alpha c} = \theta_c - \theta_a = \pi - \arcsin\frac{-3\sqrt{3}R_s-4\sqrt{3}\Delta R_a+3X_s}{\sqrt{(6R_s+9\Delta R_a)^2+(6X+\sqrt{3}\Delta R_a)^2}} - \arctan\frac{X_s}{R_s} & \Delta R_a > \frac{\sqrt{3}X_s-3R_s}{4} \end{cases} \quad (37)$$

It can be seen that when HRC occurs in phase A, while only the phase feature quantity $\theta_{\alpha a}$ has a constant value of 0, and $\theta_{\alpha b}$ and $\theta_{\alpha c}$ both change with different degrees of fault. According to (37), $\theta_{\alpha b}$ is in the range of $2\pi/3 \le \theta_{\alpha b} \le \pi$ and $\theta_{\alpha c}$ is in the range of π-arctan$X_s/R_s \le \theta_{\alpha c} \le 229\pi/180$-arctan$X_s/R_s$. Generally, synchronous reactance X_s is much greater than winding resistance R_s, and the range of $\theta_{\alpha c}$ can be from $\pi/2$ to $139\pi/180$. Similarly, the same applies when the HRC occurs in one of the other phases, and $\theta_{\alpha a}$, $\theta_{\alpha b}$, and $\theta_{\alpha c}$ only need to be replaced as $\theta_{\alpha x}$, $\theta_{\alpha y}$, and $\theta_{\alpha z}$, respectively, where x is the HRC fault phase, and x, y, z can be a, b, c, or b, c, a or c, a, b. Then, the faulty phase can be identified by calculating the phase fault characteristics $\theta_{\alpha a}$, $\theta_{\alpha b}$, and $\theta_{\alpha c}$ since only the fault phase feature quantity $\theta_{\alpha x}$ is a constant value of 0. It is a very simple and effective method for a one-phase HRC fault. When HRC occurs in two phases, such as phases A and B, there is no longer a constant characteristic quantity, and this method cannot be directly applied. In this situation, it is necessary to reanalyze its phase change and discover reliable patterns that can be used to locate the faulty phase.

Figure 5 shows the phasor diagram of DFIG stator winding with HRC in phases A and B. In Figure 5, a feature triangle is constructed both in two fault phases to help identify the pattern of phase feature quantities. It can be seen that, affected by the lagging fault phase, the current of the leading phase A leads by $-\dot{U}'_0$ a certain angle $\theta_{\alpha a}$. Similarly, the current of the lagging phase B lags behind $-\dot{U}'_0$ a certain angle $\theta_{\alpha b}$. At least one out of $\theta_{\alpha a}$ and $\theta_{\alpha b}$ is an acute angle, and one of them is the smallest of $\theta_{\alpha a}$, $\theta_{\alpha b}$ and $\theta_{\alpha c}$, which depends on the degree of fault in both phases. $-\dot{U}'_0$ is located between the leading and lagging phase currents. Then, one fault phase can be determined by finding the minimum phase characteristic quantity. By combining the phase relationship between the current of this phase and $-\dot{U}'_0$, either leading or lagging, another faulty phase can be identified. For example, when calculating, $\theta_{\alpha b}$ is the smallest of the three feature quantities and $\dot{I}'_{bs}$ leads by $-\dot{U}'_0$; then, the two fault phases are phases B and C. Otherwise, the two fault phases are phases A and B. If there are two equal minimum feature quantities, the corresponding two phases are the fault phase, as shown in Figure 5. In this situation, the phase difference between $\dot{I}'_{as}$ and $\dot{I}'_{bs}$ is $2\pi/3$, and $\theta_{\alpha a} = \theta_{\alpha b} = \pi/3$. Affected by the lagging fault phase, $\dot{I}'_{as}$ is larger than $\dot{I}'_{bs}$, and $\Delta R_a < \Delta R_b$.

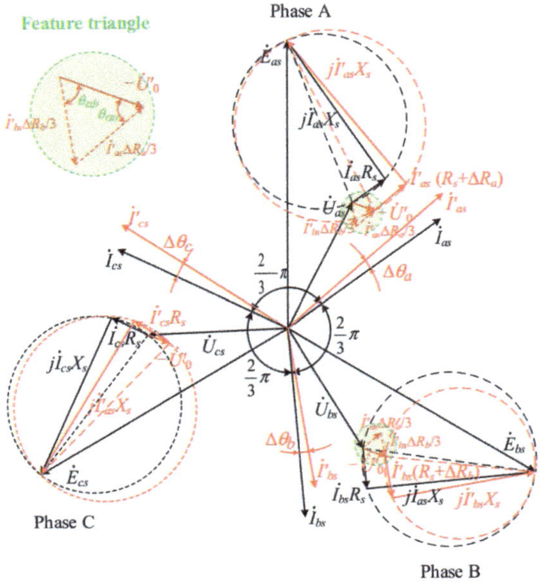

Figure 5. Phasor diagram of DFIG stator winding with HRC in phase A and B.

$\theta_{\alpha x}$ is the magnitude of the phase difference between $-\dot{U}'_0$ and $\dot{I}'_{xs}$. Through the above analysis, in order to accurately locate the faulty phase, not only the magnitude of the phase difference needs to be calculated, but also its sign needs to be determined. For the convenience of application, (34) has been modified as follows:

$$\begin{cases} \theta_{\alpha x} = |\theta_x - \alpha + \pi|, \quad S_{\alpha x} = \frac{\theta_x - \alpha + \pi}{|\theta_x - \alpha + \pi|} & \theta_x - \alpha - \pi \in [-2\pi, -\pi] \\ \theta_{\alpha x} = |\theta_x - \alpha - \pi|, \quad S_{\alpha x} = \frac{\theta_x - \alpha - \pi}{|\theta_x - \alpha - \pi|} & \theta_x - \alpha - \pi \in [-\pi, \pi] \\ \theta_{\alpha x} = |\theta_x - \alpha - 3\pi|, \quad S_{\alpha x} = \frac{\theta_x - \alpha - 3\pi}{|\theta_x - \alpha - 3\pi|} & \theta_x - \alpha - \pi \in [\pi, 2\pi] \\ \alpha \in [-\pi, \pi] \\ \theta_x \in [0, 2\pi] \\ S_{\alpha x} = 1, \qquad \text{if } \theta_{\alpha x} = 0 \end{cases} \tag{38}$$

where $x = a, b,$ or c, α is the extracted initial phase of $\dot{U}'_0$ and $\theta_{\alpha x}$ is the extracted initial phase of $\dot{I}_{xs}$. When the HRC occurs in one phase, the additional resistance ΔR_x can be obtained using (32) as follows:

$$\Delta R_x = \frac{3U_0}{I_{xs}} \tag{39}$$

When the HRC occurs in two phases, the additional resistance ΔR_x and ΔR_y can be obtained using the feature triangle as follows:

$$\begin{cases} \Delta R_x = \frac{3U_0 \sin \theta_{\alpha y}}{I_{xs} \sin(\pi - \theta_{\alpha x} - \theta_{\alpha y})} \\ \Delta R_y = \frac{3U_0 \sin \theta_{\alpha x}}{I_{ys} \sin(\pi - \theta_{\alpha x} - \theta_{\alpha y})} \end{cases} \tag{40}$$

Then, the flowchart of the HRC diagnosis process is provided in Figure 6. The online HRC fault diagnosis process is as follows:

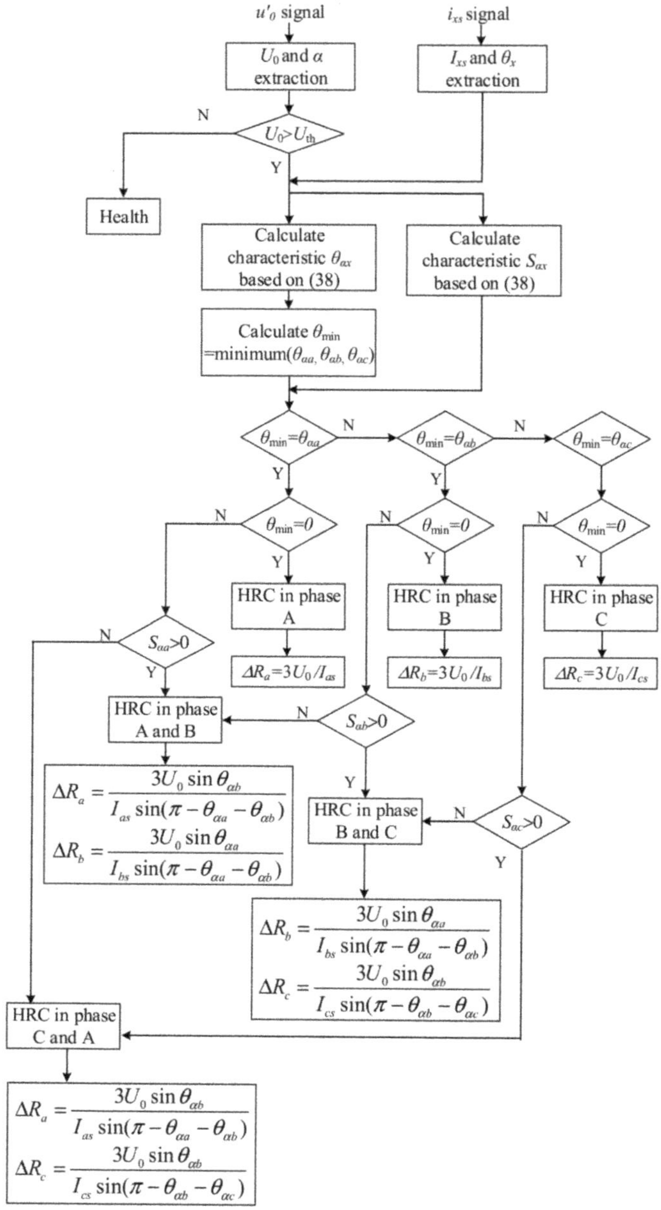

Figure 6. Flow chart of HRC diagnosis process.

Step 1: HRC fault detection. The fundamental components of u'_0 and i_{xs} are extracted online. Then, once the amplitude U_0 is larger than 0, the HRC fault can be detected.

Step 2: Fault phase location. After detecting the occurrence of HRC, HRC fault characteristics $\theta_{\alpha x}$ and $S_{\alpha x}$ are calculated based on (38). Then, according to the minimum of $\theta_{\alpha x}$ and combined with the sign of $S_{\alpha x}$, the fault phase can be located.

Step 3: Fault severity estimation. After locating the fault phase, the additional resistance ΔR_x can be calculated based on (39) or (40).

4. Simulations

In order to verify the effectiveness of the proposed method and ensure the reliability of the simulation results, the complex electromagnetic environment of the DFIG should be simulated realistically. Then, considering the impact of complex electromagnetic fields on fault diagnosis, a finite element (FE) model of DFIG can be established alongside the external circuit model based on the field-circuit coupling method, as shown in Figure 7. The main parameters of the FE model are shown in Table 1.

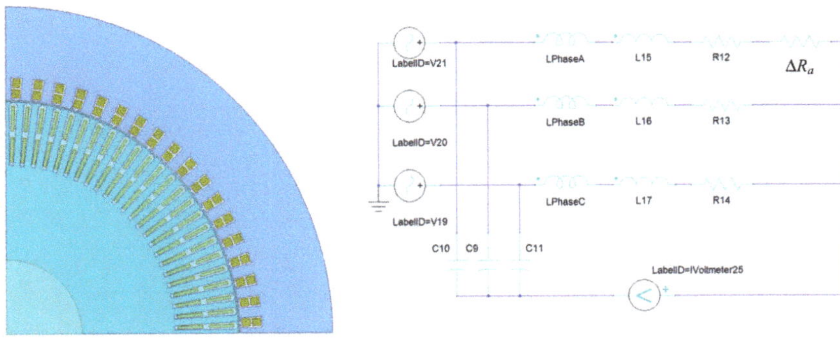

(a) (b)

Figure 7. Field-circuit coupling model of DFIG. (**a**) FE simulation model; (**b**) External circuit.

Table 1. FE Model parameters of DFIG.

Item	Value	Item	Value
Rated power [kW]	1500	Rated speed [rpm]	1750
Stator outer diameter [mm]	860	Stack length [mm]	780
Stator inner diameter [mm]	615	Pole-pair number	2
Thickness of stator yoke [mm]	93.5	Number of stator slot	72
Tooth width of stator [mm]	14	Number of rotor slot	96
Rotor outer diameter [mm]	611.4	Number of stator winding layers	2
Rotor inner diameter [mm]	200	Number of rotor winding layers	2
Thickness of rotor yoke [mm]	122.5	Stator coil pitch	16
Tooth width of rotor [mm]	9.3	Rotor coil pitch	20

Figure 8 shows the waveform and harmonic components of the voltage u'_0 in the healthy DFIG. It can be seen that, in the healthy DFIG, the main component of u'_0 is the third harmonic component, and the fundamental component is very small and almost non-existent. As shown in Figure 7b, an additional resistor $\Delta R = 0.2\ \Omega$ is connected in series with A-phase winding, and the resistance of A-phase winding in a healthy state is 0.2 Ω. Then, the simulation results are shown in Figure 9. Compared with Figure 8, the fundamental component appears in the voltage u'_0, where $U_0 = 4.79255$ V and $\alpha = -20.7939°$, as shown in Figure 9a. From Figure 9b, it can be seen that $\theta_a = 159.358°$ and I_{as} =73.1161 A before $\theta_{\alpha a} = |\theta_a - \alpha - \pi| = 0$ is obtained. According to the proposed diagnosis method, the HRC fault is in A-phase winding, and according to (39), the additional resistance ΔR_a can be calculated as $\Delta R_a = 3U_0/I_{as} = 0.19664\ \Omega$. So, the proposed diagnosis method is effective for HRC in the one-phase stator winding of the DFIG.

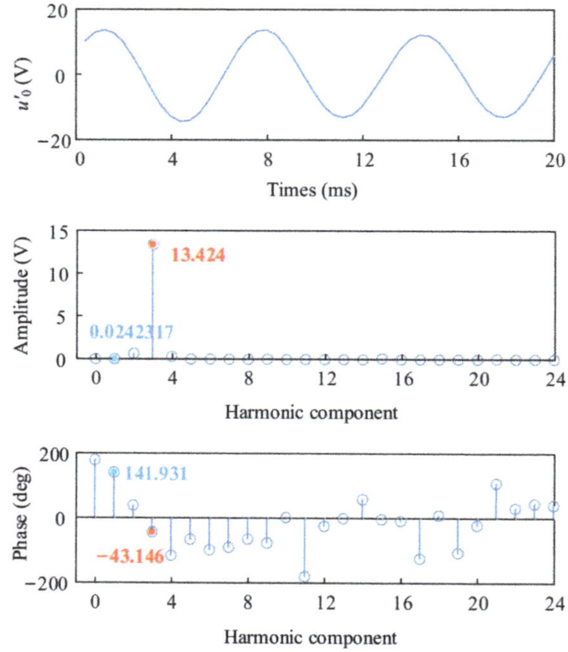

Figure 8. u'_0 of the healthy DFIG.

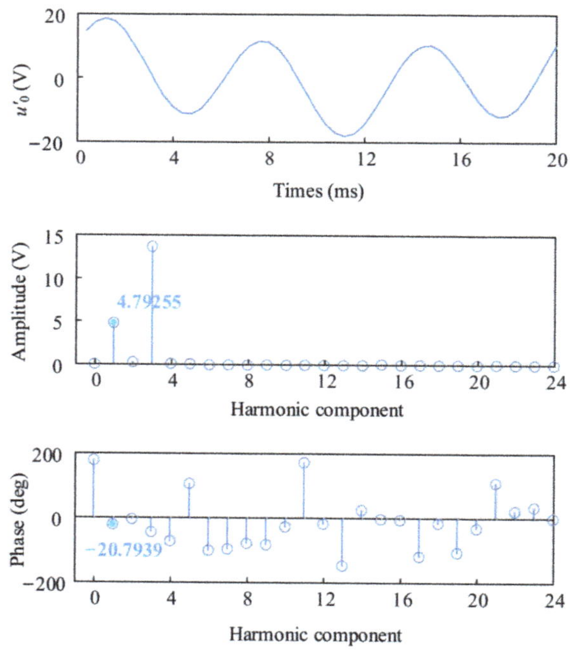

(a)

Figure 9. *Cont.*

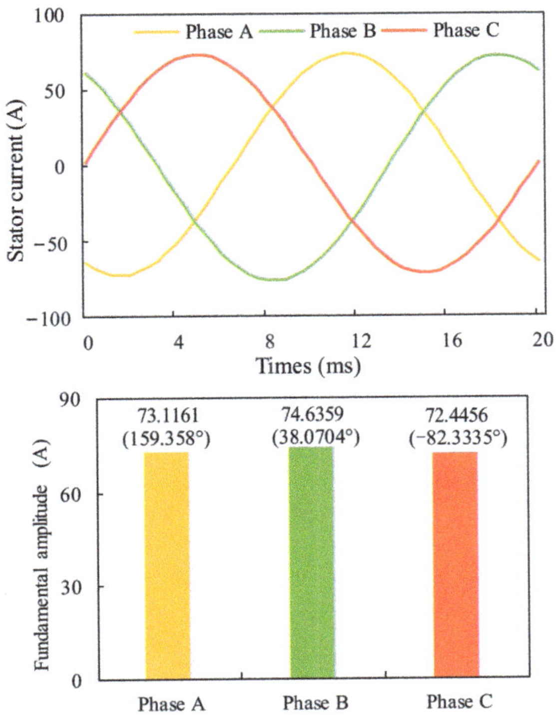

(b)

Figure 9. u'_0 and stator currents of the DFIG with HRC in phase A. (**a**) u'_0; (**b**) Stator currents.

Figure 10 shows the simulation results when HRC occurs in phase A and phase B, where $\Delta R_a = \Delta R_b = 0.2\ \Omega$. From Figure 10, it is shown that $U_0 = 4.80292$ V, $\alpha = -81.817°$, $I_{as} = 71.8809$ A, $I_{bs} = 74.1309$ A, $I_{cs} = 73.2912$ A, $\theta_a = 159.812°$, $\theta_b = 40.0469°$, and $\theta_c = -81.5924°$. According to (38), $\theta_{\alpha a} = |\theta_a - \alpha - \pi| = 61.629°$, $S_{\alpha a} = 1$, $\theta_{\alpha b} = |\theta_b - \alpha - \pi| = 58.1361°$, $S_{\alpha b} = -1$, $\theta_{\alpha c} = |\theta_c - \alpha - \pi| = 179.7754°$, and $S_{\alpha c} = -1$. Then, $\theta_{min} = \theta_{\alpha b}$, and $S_{\alpha b} < 0$. According to Figure 6, the HRC fault is in phases A and B. According to (40), it is shown that $\Delta R_a = 0.19612\ \Omega$ and $\Delta R_b = 0.19702\ \Omega$. It can be seen that the error of the proposed diagnosis method is only 1.49%, which is a high accuracy.

Figure 11 shows the simulation results when HRC occurs in phase A and phase B, and the degree of fault between phase A and phase B is different, where $\Delta R_a = 0.2\ \Omega$, $\Delta R_b = 0.5\ \Omega$. From Figure 11, it is shown that $U_0 = 10.7089$ V, $\alpha = -114.52°$, $I_{as} = 69.9473$ A, $I_{bs} = 73.3283$ A, $I_{cs} = 74.6458$ A, $\theta_a = 160.449°$, $\theta_b = 43.1946°$, and $\theta_c = -80.3946°$. It can be seen that U_0 in Figure 11 is greater than that in Figure 10 since the fault degree in Figure 11 is greater than that in Figure 10. According to (38), $\theta_{\alpha a} = |\theta_a - \alpha - \pi| = 94.969°$, $S_{\alpha a} = 1$, $\theta_{\alpha b} = |\theta_b - \alpha - \pi| = 22.2854°$, $S_{\alpha b} = -1$, $\theta_{\alpha c} = |\theta_c - \alpha - \pi| = 145.8746°$, $S_{\alpha c} = -1$. Then, $\theta_{min} = \theta_{\alpha b}$ and $S_{\alpha b} < 0$. According to Figure 6, the HRC fault is in phases A and B. According to (40), it has $\Delta R_a = 0.1959\ \Omega$ and $\Delta R_b = 0.5147\ \Omega$. It can be seen that the HRC in phases A and B can be accurately diagnosed.

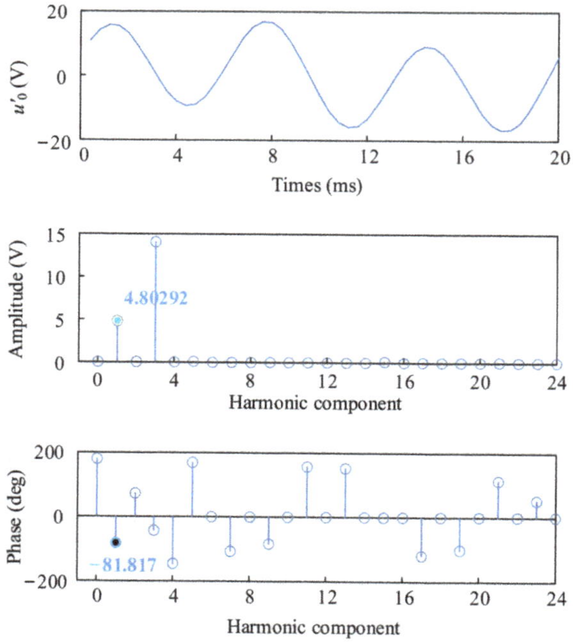

(a)

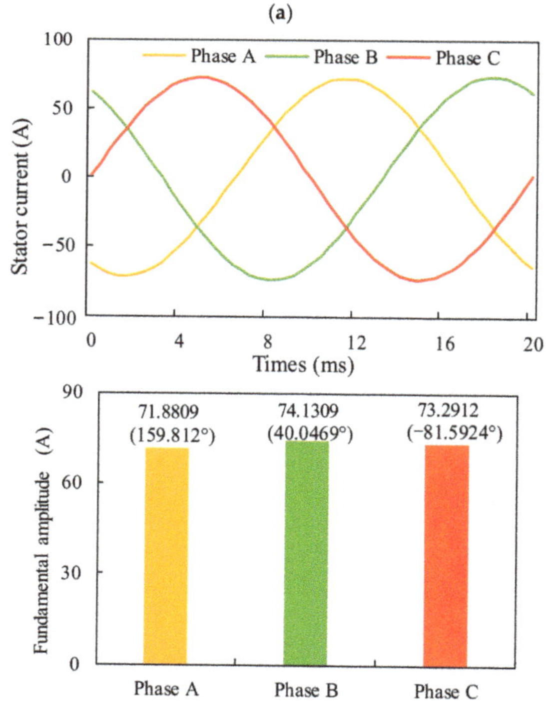

(b)

Figure 10. u'_0 and stator currents of the DFIG with HRC in phase A and B. ($\Delta R_a = \Delta R_b = 0.2\ \Omega$) (**a**) u'_0; (**b**) Stator currents.

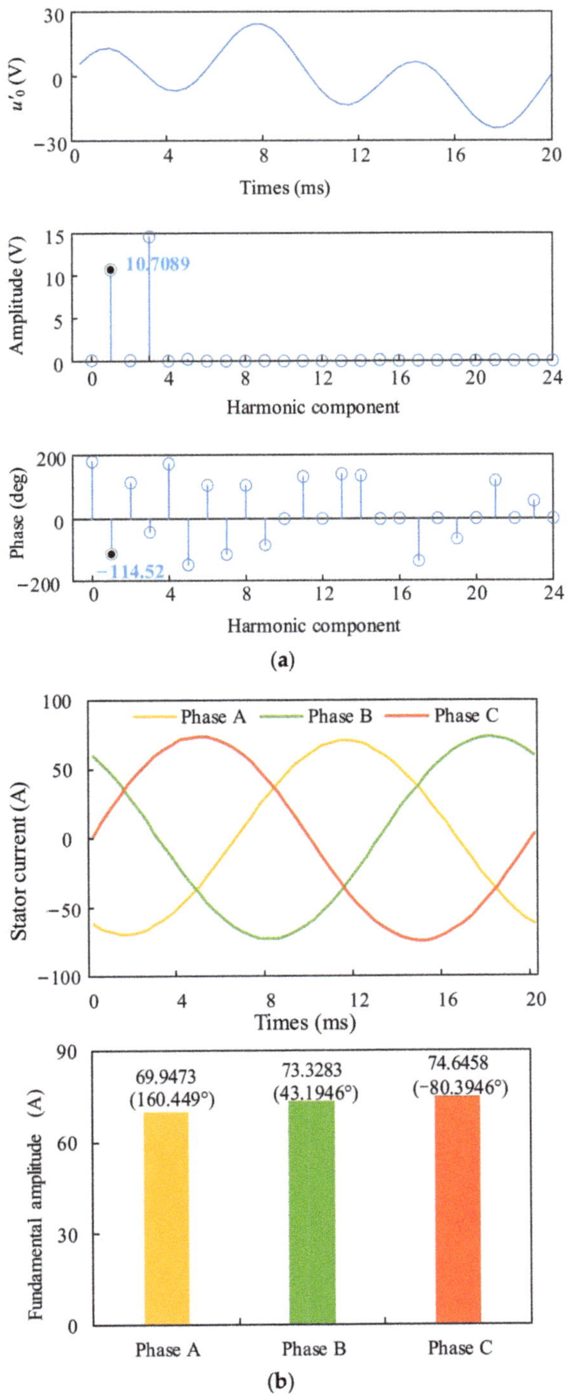

Figure 11. u'_0 and stator currents of the DFIG with HRC in phase A and B. (ΔR_a = 0.2 Ω, ΔR_b = 0.5 Ω) (**a**) u'_0; (**b**) Stator currents.

Figure 12 shows the simulation results when HRC occurs in phase A and phase C, and the degree of fault between phase A and phase C is different, where ΔR_a = 0.2 Ω, ΔR_c = 0.5 Ω. From Figure 12, it is shown that U_0 = 9.96105 V, α = 77.967°, I_{as} = 75.3756 A, I_{bs} = 71.5241 A, I_{cs} = 71.1704 A, θ_a = 161.3°, θ_b = 39.1857°, and θ_c = −77.043°. According to (38), $\theta_{\alpha a} = |\theta_a - \alpha - \pi|$ = 96.667°, $S_{\alpha a} = -1$, $\theta_{\alpha b} = |\theta_b - \alpha + \pi|$ = 141.2187°, $S_{\alpha b} = 1$, $\theta_{\alpha c} = |\theta_c - \alpha - \pi|$ = 24.99°, $S_{\alpha c} = 1$. Then, $\theta_{min} = \theta_{\alpha c}$ and $S_{\alpha b} > 0$. According to Figure 6, the HRC fault is in phases A and C. According to (40), it has ΔR_a = 0.1968 Ω and ΔR_b = 0.4875 Ω. The HRC in phases A and B can also be accurately diagnosed. So, the proposed diagnosis method is effective for HRC in two-phase stator windings of the DFIG.

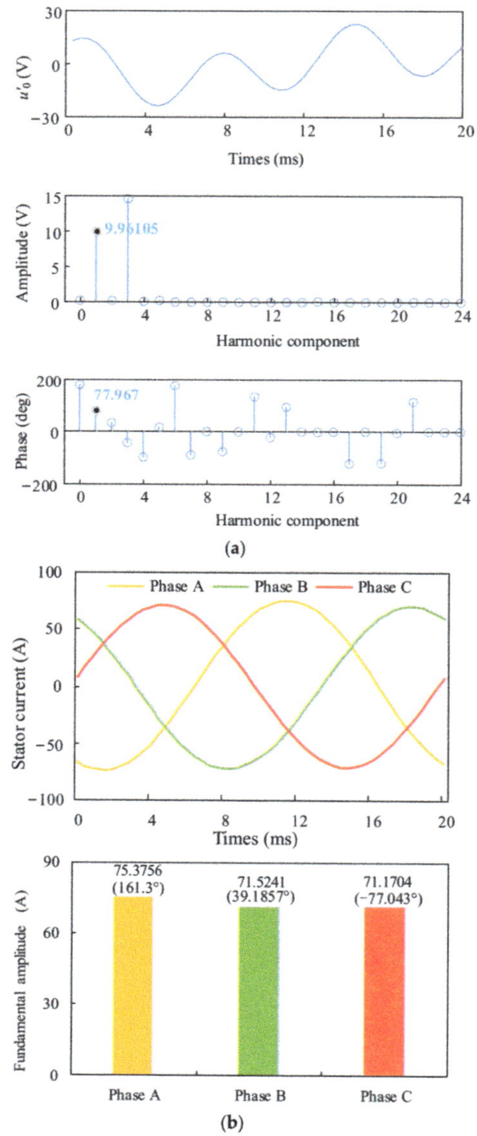

Figure 12. u'_0 and stator currents of the DFIG with HRC in phase A and C. (ΔR_a = 0.2 Ω, ΔR_c = 0.5 Ω) (**a**) u'_0; (**b**) Stator currents.

5. Discussion of Results

Based on the above simulation analysis, it can be seen that the method proposed in this article cannot only achieve an accurate fault degree evaluation and fault location of single-phase faults but can also accurately locate the fault phase of two-phase faults, and accurately evaluate the fault degree of each phase. The methods in [10,11] can only achieve accurate fault diagnosis for single-phase faults, and, for two-phase faults, only an approximate fault degree could be estimated, where the fault degree judged by this method was the same for two-phase faults, and it could not evaluate the faults of each phase separately. Meanwhile, from the processing flow, it can be seen that the proposed method was relatively simple compared to other signal processing-based methods [12,13]. Moreover, its diagnostic process does not rely on precise DFIG models like [4,14], but it has a stronger anti-interference ability. And compared to knowledge-based methods [16,17], it does not need to rely on a large number of samples, it is not affected by the training model, and can achieve high-precision fault diagnosis at a small cost.

The above analysis is based on a simulation, and there may be some differences between the experimental results and simulation results. Firstly, the inherent asymmetry of the generator itself can cause the average value of the sliding window to be greater than zero under normal circumstances. Therefore, in practical situations, appropriate thresholds should be selected to avoid the misdiagnosis caused by inherent asymmetry. Secondly, in practice, there may be measurement interferences in voltage sensors and current sensors, which leads to certain burrs in the actual phase information and affects the accuracy of fault assessment, but it does not have a significant impact on fault location.

6. Conclusions and Future Work

This paper proposes a new method for the online diagnosis of DFIGs stator winding HRC faults using artificial neutral points constructed using parallel capacitor banks. Based on the analysis of the mathematical model of DFIG with HRC in stator winding, the fault feature quantities are defined, and the HRC fault is diagnosed based on the fault feature quantities. Finally, the simulations are conducted to verify the effectiveness and correctness of the theoretical analysis. The research content of this paper is summarized as follows:

(1) The interference and false alarms caused by power grid imbalance can be eliminated through the method of constructing an artificial neutral point, as proposed in this paper.
(2) The proposed method can accurately locate the faulty phase and evaluate the degree of the fault in the case of single-phase faults, with an evaluation accuracy of over 98%.
(3) In the case of faults occurring in two phases, regardless of whether the faults are the same or not, the proposed method can accurately locate the faulty phase and evaluate the degree of fault in each phase, with an evaluation accuracy of over 97%.

The proposed method is based on the assumption of an excellent control performance. The next step, the impact of different control methods and control performance, needs to be analytically researched. Meanwhile, considering the feasibility of the application of this method in rotor winding fault diagnosis, a classification method for diagnosing HRC faults in stator and rotor windings should be studied.

Author Contributions: Conceptualization, W.D. and Y.J. (Yulong Jin); methodology, W.D., Y.J. (Yulong Jin), X.W., Y.Y. and Y.J. (Yongjiang Jiang); validation, X.W., Y.Y. and Y.J. (Yongjiang Jiang); formal analysis, W.D. and Y.J. (Yulong Jin); investigation, Y.J. (Yongjiang Jiang) and X.W.; writing—original draft preparation, W.D. and Y.J. (Yulong Jin); writing—review and editing, X.W., Y.Y. and Y.J. (Yongjiang Jiang); supervision, Y.J. (Yulong Jin) and X.W; project administration, Y.J. (Yulong Jin); funding acquisition, Y.Y. All authors have read and agreed to the published version of the manuscript.

Funding: This research was funded by the Research and Application of Key Technologies for Centralized Control Operation and Maintenance of New Energy, grant number 524609220028.

Data Availability Statement: Data are contained within the article.

Conflicts of Interest: The authors declare no conflict of interest.

References

1. Hang, J.; Xia, M.; Ding, S.; Li, Y.; Sun, L. Research on vector control strategy of surface-mounted permanent magnet synchronous machine drive system with high-resistance connection. *IEEE Trans. Power Electron.* **2020**, *35*, 2023–2033. [CrossRef]
2. Hang, J.; Ren, X.; Tang, C.; Tong, M. Fault-tolerant control strategy for five-phase PMSM drive system with high-resistance connection. *IEEE Trans. Transp. Electrific.* **2021**, *7*, 1390–1400. [CrossRef]
3. Xu, Z.; Din, Z.; Jiang, Y.; Cheema, K.M.; Milyani, A.H.; Alghamdi, S. High-resistance connection diagnosis considering current closed-loop effect for permanent magnet machine. *Front. Energy Res.* **2022**, *10*, 933246. [CrossRef]
4. Gonçalves, P.F.C.; Cruz, S.M.A.; Mendes, A.M.S. Online diagnostic method for the detection of high-resistance connections and open-phase faults in six-phase PMSM drives. *IEEE Trans. Ind. Appl.* **2022**, *58*, 345–355. [CrossRef]
5. Rossi, C.; Gritli, Y.; Pilati, A.; Rizzoli, G.; Tani, A.; Casadei, D. High resistance fault-detection and fault-tolerance for asymmetrical six-phase surface-mounted AC permanent magnet synchronous motor drives. *Energies* **2020**, *13*, 3089. [CrossRef]
6. Zamudio-Ramirez, I.; Antonino-Daviu, J.A.; Osornio-Rios, R.A.; de Jesus Romero-Troncoso, R.; Razik, H. Detection of winding asymmetries in wound-rotor induction motors via transient analysis of the external magnetic field. *IEEE Trans. Ind. Electron.* **2020**, *67*, 5050–5059.
7. Zamudio-Ramirez, I.; Antonino-Daviu, J.A.; Osornio-Rios, R.A.; Dunai, L. Tracking of high-order stray-flux harmonics under starting for the detection of winding asymmetries in wound-rotor induction motors. *IEEE Trans. Ind. Electron.* **2022**, *69*, 8463–8471. [CrossRef]
8. Zamudio-Ramirez, I.; Antonino-Daviu, J.; Osornio-Rios, R.A.; Dunai, L.; Quijano-Lopez, A.; Fuster-Roig, V. Detection of stator asymmetries in wound rotor induction motors through the advanced analysis of rotor currents. In Proceedings of the 2022 IEEE Energy Conversion Congress and Exposition (ECCE), Detroit, MI, USA, 9–13 October 2022.
9. Martínez, M.E.I.; Antonino-Daviu, J.A.; de Córdoba, P.F.; Conejero, J.A.; Dunai, L. Automatic classification of winding asymmetries in wound rotor induction motors based on bicoherence and fuzzy c-means algorithms of stray flux signals. *IEEE Trans. Ind. Appl.* **2021**, *57*, 5876–5886. [CrossRef]
10. Wang, H.; Lu, S.; Qian, G.; Ding, J.; Liu, Y.; Wang, Q. A two-step strategy for online fault detection of high-resistance connection in bldc motor. *IEEE Trans. Power Electron.* **2020**, *35*, 3043–3053. [CrossRef]
11. Hang, J.; Yan, D.; Xia, M.; Ding, S.; Wang, Q. Quantitative fault severity estimation for high-resistance connection in PMSM drive system. *IEEE Access* **2019**, *7*, 26855–26866. [CrossRef]
12. Hu, R.; Wang, J.; Mills, A.R.; Chong, E.; Sun, Z. Detection and classification of turn fault and high resistance connection fault in permanent magnet machines based on zero sequence voltage. *IEEE Trans. Power Electron.* **2020**, *35*, 1922–1933. [CrossRef]
13. Hang, J.; Wu, H.; Ding, S.; Hua, W.; Wang, Q. A DC-flux-injection method for fault diagnosis of high-resistance connection in direct-torque-controlled PMSM drive system. *IEEE Trans. Power Electron.* **2020**, *35*, 3029–3042. [CrossRef]
14. Hang, J.; Zhang, J.; Ding, S.; Huang, Y.; Wang, Q. A model-based strategy with robust parameter mismatch for online HRC diagnosis and location in PMSM drive system. *IEEE Trans. Power Electron.* **2020**, *35*, 10917–10929. [CrossRef]
15. Kommuri, S.K.; Park, Y.; Bin Lee, S. High-resistance fault control in permanent magnet synchronous motors. *IEEE/ASME Trans. Mech.* **2019**, *25*, 271–281. [CrossRef]
16. Lee, H.; Jeong, H.; Koo, G.; Ban, J.; Kim, S.W. Attention Recurrent Neural Network-Based Severity Estimation Method for Interturn Short-Circuit Fault in Permanent Magnet Synchronous Machines. *IEEE Trans. Ind. Electron.* **2021**, *68*, 3445–3453. [CrossRef]
17. Li, Y.; Wang, R.; Mao, R.; Zhang, Y.; Zhu, K.; Li, Y.; Zhang, J. A Fault Diagnosis Method Based on an Improved Deep Q-Network for the Inter-Turn Short Circuits of a Permanent Magnet Synchronous Motor. *IEEE Trans. Ind. Electron.* **2023**, *in press*. [CrossRef]
18. Faiz, J.; Abadi, M.B.; Cruz, S.M.; Moosavi, S.M.M. Comparison of rotor electrical fault indices owing to inter-turn short circuit and unbalanced resistance in doubly-fed induction generator. *IET Electr. Power App.* **2019**, *13*, 235–242.
19. Battulga, B.; Shaikh, M.F.; Bin Lee, S.; Osama, M. Automated identification of failures in doubly-fed induction generators for wind turbine applications. *IEEE Trans. Ind. Appl.* **2023**, *59*, 4454–4463. [CrossRef]
20. Ibrahim, R.K.; Watson, S.J.; Djurović, S.; Crabtree, C.J. An effective approach for rotor electrical asymmetry detection in wind turbine DFIGs. *IEEE Trans. Ind. Electron.* **2018**, *65*, 8872–8881. [CrossRef]
21. Fu, Y.; Ren, Z.; Wei, S.; Xu, Y.; Li, F. Using flux linkage difference vector in early inter-turn short circuit detection for the windings of offshore wind DFIGs. *IEEE Trans. Energy Convers.* **2021**, *36*, 3007–3015. [CrossRef]

energies

Article

Fixed-Time Sliding Mode Control for Linear Motor Traction Systems with Prescribed Performance

Chunguang Yang [1,†], Guanyang Hu [2,†], Qichao Song [1,*], Yachao Wang [3] and Weilin Yang [2]

1 School of Control Technology, Wuxi Institute of Technology, Wuxi 214121, China
2 School of Internet of Things Engineering, Jiangnan University, Wuxi 214122, China
3 State Grid Jibei Marketing Service Center, No. 1, Dizangnanxiang Road, Beijing 100000, China
* Correspondence: yang20200927@163.com
† These authors contributed equally to this work.

Abstract: In this research, we propose a fixed-time sliding mode controller using a prescribed performance control approach to address the speed tracking problem in linear motor traction systems, which are powered by high-power permanent magnet linear synchronous motors (PMLSMs). Initially, to tackle the issue of the convergence time and dynamic response associated with traditional finite-time sliding mode controllers, we introduce a fixed-time sliding mode controller. This controller guarantees that the system state converges to the origin within a specified upper time limit. Subsequently, to enhance the dynamic response of the PMLSM and minimize speed errors, we integrate the prescribed performance control strategy with a fixed-time sliding mode controller. This effectively limits the motor's speed error within the predefined function boundaries, reduces system overshoo, and mitigates system jitter to a certain degree. Finally, simulation results are presented to validate that the proposed control strategy significantly enhances precision of speed tracking in PMLSMs.

Keywords: fixed-time sliding mode controller; prescribed performance; linear motor traction systems; permanent magnet linear synchronous motors

Citation: Yang, C.; Hu, G.; Song, Q.; Wang, Y.; Yang, W. Fixed-Time Sliding Mode Control for Linear Motor Traction Systems with Prescribed Performance. *Energies* **2024**, *17*, 952. https://doi.org/10.3390/en17040952

Academic Editor: Adolfo Dannier

Received: 9 January 2024
Revised: 14 February 2024
Accepted: 17 February 2024
Published: 18 February 2024

1. Introduction

With the rapid growth in the global economy and population, traditional transportation methods are facing increasing pressure, particular in the form of urban traffic congestion. Therefore, the development of new high-efficiency, environmentally friendly vehicles, and transportation systems has become an important direction for research and development in the global transportation field. It is against this background that high-power linear traction systems came into being. They use the force generated by high-intensity electromagnetic fields to drive objects to move on guide rails, thereby achieving high-speed, low-energy, and low-noise transportation [1,2]. Compared with traditional fuel-driven systems and mechanical rotating transmission systems, high-power linear traction systems are have the characteristics of highly efficient, fast, strong, and environmentally friendly, so they are widely used for long-distance and so high-speed transportation.

Generally, linear traction systems are powered by a high-power permanent magnet linear synchronous motor (PMLSM), which possess many advantages, e.g., fast speed response, accurate fast positioning, and zero-transmission characteristics. Hence, these systems offer a low mechanical loss, large thrust density, and a fast dynamic response [3,4]. Likewise, they are widely employed in industrial production and rail transit systems, e.g., industrial robots, computer numerical control machine tools, and maglev trains etc. [5,6].

A PMLSM's control system is a complex nonlinear system and therefore requires advanced control strategies [7–9]. A cascade control arrangement is commonly utilized in the control systems of PMLSMs. This structure comprises an outer control loop for speed, and an inner control loop for current, intended to follow the specified speed and the current references, respectively. The former plays an important role in the control system and requires advanced control strategies to achieve precise speed tracking. The commonly used control methods for the inner loop include proportional-integral (PI) control [10], current hysteresis control, sliding mode control (SMC), and some other intelligent control methods [11–13].

In an actual control system, PMLSMs are susceptible to various external disturbances, such as mechanical noise, vibration, and model mismatch caused by changes in the parameters of the equipment itself during long-term operation, which will lead to a deterioration in the control system's performance and affect the motor's tracking accuracy. PI controllers have been widely used in PMLSMs. However, they are vulnerable to external interferences to some extent. Hence, various approaches to improving the anti-disturbance capability of a PMLSM controller have been proposed. Specifically, typical designs often rely on the outer control loop, namely, the speed control loop, within the cascaded structure. Among various methodologies, SMC has attracted significant attention forom numerous researchers investigating PMLSM control due to its simplicity and robustness in the face of disturbances. In [14], a sliding mode controller was developed to replace the PI controller, demonstrating a superior control performance. Generally, the SMC method involves selecting a linear sliding surface. Once the system reaches the sliding mode surface, the speed tracking error gradually diminishes to zero. The convergence speed is adjustable by modifying the sliding mode surface parameters, although the state tracking error does not converge to zero within a finite time. Recent studies have proposed a terminal sliding mode control (TSMC) strategy to address this limitation. In TSMC, a nonlinear function is introduced into the sliding mode surface, constructing a terminal sliding mode surface. As a result, after reaching the sliding mode surface, the system's tracking error converges to zero within a finite time [15,16].

To address the issue of the slow convergence in traditional sliding mode control algorithms, scholars have introduced a fixed-time sliding mode control algorithm. This approach ensures system convergence within a predetermined time, independent of the initial system state [17–19]. In [20], researchers proposed a fixed-time distributed sliding mode control method to attain formation control in fractional-order multi-agent systems. Numerical simulations showcase its superior convergence rate compared to the finite-time sliding mode control strategy. An innovative fixed-time nonsingular fast terminal sliding mode control method was presented in [21] to achieve rapid stability and robust control for second-order nonlinear systems. This method not only guarantees fixed-time convergence but also mitigates singularity issues present in conventional terminal sliding mode surfaces. Addressing speed regulation system control for permanent magnet synchronous motors, ref. [22] introduced an integral fixed-time sliding mode control algorithm with disturbance estimation compensation. Rigorous Lyapunov function analyses established that the speed tracking error converges to zero within a fixed time. Comparative results from numerical experiments confirmed the effectiveness and superiority of the integral fixed-time sliding mode control method. The collective findings suggest that fixed-time sliding mode controller accelerates system state convergence and enhances the dynamic performance compared to the finite-time sliding mode controllers.

In servo control systems, beyond considering the system's dynamic performance, it is crucial to assess the steady-state speed error to significantly improve performance. In prescribed performance control (PPC), pre-setting controllers are adopted to meet system performance targets and requirements. This approach has proven advantageous for enhancing system's stability, control accuracy, and adaptability, and so it is widely used in servo control systems [23,24], vehicle systems [25], and marine systems [26]. To precisely regulate the tracking performance of servo systems, including the overshoot, convergence speed, and steady-state error, a composite finite-time control scheme with prescribed performance for speed regulation in permanent magnet synchronous motors has been explored in [27]. Initially, prescribed transient and steady-state performance constraints are considered by using PPC. Subsequently, a composite finite-time speed controller is devised based on a feed-forward compensated disturbance observer, and the finite-time stability of the closed-loop system was meticulously analyzed. Finally, the control scheme's effectiveness is validated through numerical simulations. However, the PPC method is susceptible to external perturbations, and when a sudden change in motor load occurs, the speed error may break through prescribed boundaries, affecting the control performance. Consequently, PPC is often integrated with robust control algorithms such as sliding mode control (SMC) strategy.

Inspired by the above research, in this study,we introduce a fixed-time sliding mode controller designed under the framework of prescribed performance control (PPC-FTSMC). The objective is to achieve precise speed control, enhance the system's dynamic performance, and reduce steady-state errors. Employing field orientation control (FOC) in the control structure, PPC-FTSMC is devised for the speed loop, complemented by two PI controllers within the current loops. MATLAB simulations are presented to demonstrate the robustness and dynamic efficacy of the proposed strategy. The key contributions of this work are outlined as follows:

- PPC is used to restrict the tracking errors, leading to enhancements in the dynamic response, mitigation of overshoot, and a reduction in tracking errors.
- FTSMC contributes to an increased system robustness and accelerates the convergence time of speed errors.
- The fixed-time stability of both the sliding mode surface and the system states under the composite control scheme proposed is substantiated through Lyapunov stability theory.

The remainder of this paper is organized as follows. Section 2 introduces some important lemmas and describes the mathematical model of PMLSMs. Likewise, Section 3 states a PPC-FTSMC method, and it is applied to the inner loop. Lastly the results from simulations and corresponding experiments are provided in Section 4, and the work is concluded in Section 5.

2. Problem Description

2.1. Mathematical Modelling of PMLSMs

Before exploring the mathematical model of the PMLSM, it is crucial to set certain assumptions. These assumptions involve disregarding the saturation of the motor core, excluding the losses caused by eddy currents and hysteresis in the motor, and assuming that the three-phase current waveform follows an ideal sine wave pattern.

We can establish the electromagnetic thrust equation can be written as follows:

$$F_e = K_f i_q = M\dot{v} + Bv + d, \tag{1}$$

$$d = F_L + F_f + F_r, \tag{2}$$

and F_e is electromagnetic thrust. Moreover, $K_f = \frac{3}{2}\frac{\pi}{\tau}n\psi_f$ is the thrust coefficient, i_q are the stator currents on q axis, τ is polar distance, B_v is viscous friction coefficient, v represent the speed for PMLSM, n is number of pole pairs, and M is mover mass. d represents the disturbance, including load disturbance F_L, friction between the motor and the guide rail F_f and thrust fluctuations caused by end effects F_r.

For ease of expression, its dynamic equation can be rewritten as:

$$\frac{dv}{dt} = A_m v + B_m iq + D \tag{3}$$

where $A_m = -B/M$, $B_m = K_f/M$, $D = -d/M$.

2.2. Some Lemmas and Assumptions

Lemma 1 ([20]). *If there is a continuous radially bounded function $V : R^n \to R_+ \cup \{0\}$ satisfied :*

(1) $V(x) = 0 \Leftrightarrow x = 0$.

(2) For any positive constant α, β, γ_1 and γ_2 , if a nonnegative function satisfies $\dot{V}(x) \leq -\gamma_1 V^{\alpha}(x) - \gamma_2 V^{\beta}(x)$, where $0 < \alpha < 1, \beta > 1$. The system will converge in a fix time, and its convergence time is :

$$T \leq T_{\max} = \frac{1}{\gamma_1(1-\alpha)} + \frac{1}{\gamma_2(\beta - 1)} \tag{4}$$

Assumption 1: *The following conditions must be met in prescribed performance function:*
(1) $\sigma(t)$ is a monotonically decreasing positive function
(2) $\lim_{t \to \infty} \sigma(t) = \sigma_\infty > 0$.

Assumption 2: *In the motor dynamics Equation* (3), *the concentrated disturbance D of the system is bounded, there exists a constant $l > 0$* that $|D| \leq l$.

3. PPC-FTSMC Schemes for the Velocity Control Loop

3.1. Prescribed Performance Control

The dynamic characteristics of PMLSM can be improved by utilizing the PPC, which restricts the actual output speed of PMLSM within a close range to the reference speed and confines the maximum overshoot within a pre-determined threshold [28]. One possible choice for the prescribed performance function is:

$$\sigma(t) = (\sigma_0 - \sigma_\infty)\exp(-\lambda t) + \sigma_\infty \tag{5}$$

where σ_0, σ_∞, and λ are positive constants and $\sigma_0 > \sigma_\infty$. σ_∞ represents the bound of error, λ represents the convergence rate of the dynamic response of the system. Theoretically, the larger σ_0 and σ_∞ are, the smaller the residual set of the system tracking error is. In addition, increasing λ means that the tracking error convergence rate can be increased.

The error $e(t)$ satisfied:

$$\begin{cases} -\delta\sigma(t) < e(t) < \sigma(t), e(0) \geq 0 \\ -\sigma(t) < e(t) < \delta\sigma(t), e(0) < 0 \end{cases} \tag{6}$$

where $0 \leq \delta \leq 1$.

By introducing a smooth increasing function $\Psi(\varepsilon)$, where ε is the transformed error. The error inequality (6) of the system can be changed to the transformation error form.

$$\begin{cases} -\delta < \Psi(\varepsilon) < 1, e(0) \geq 0 \\ -1 < \Psi(\varepsilon) < \delta, e(0) < 0 \end{cases} \tag{7}$$

when $e(0) \geq 0$, $\lim_{\varepsilon \to -\infty} \Psi(\varepsilon) = -\delta$ and $\lim_{\varepsilon \to +\infty} \Psi(\varepsilon) = 1$, while $e(0) < 0$, $\lim_{\varepsilon \to -\infty} \Psi(\varepsilon) = -1$ and $\lim_{\varepsilon \to +\infty} \Psi(\varepsilon) = \delta$.

It is worth noting that when $e(0) = 0$, δ cannot be zero because it will cause ε to be infinite. So $\Psi(\varepsilon)$ can be described as follows:

$$\Psi(\varepsilon) = \begin{cases} \dfrac{\exp(\varepsilon) - \delta\exp(-\varepsilon)}{\exp(\varepsilon) + \exp(-\varepsilon)}, e(0) \geq 0 \\ \dfrac{\delta\exp(\varepsilon) - \exp(-\varepsilon)}{\exp(\varepsilon) + \exp(-\varepsilon)}, e(0) < 0 \end{cases} \tag{8}$$

According to the Formula (8),

$$e(t) = \sigma(t)\Psi(\varepsilon) \tag{9}$$

Since $\Psi(\varepsilon)$ is strictly monotonically increasing its inverse function must exist

$$\varepsilon = \Psi^{-1}[e(t)/\sigma(t)] \tag{10}$$

As indicated in the literature [29], when ε remains within bounds, Equation (10) remains valid. The tracking error e within the system is transformed into an unbounded conversion error, denoted as ε through PPC. Consequently, the original boundary constraints on tracking error are imposed, converting error control into the stability control of the system relative to ε. The expression for ε is given by:

$$\varepsilon = \Psi^{-1}\left(\frac{e(t)}{\sigma(t)}\right) = \begin{cases} \frac{1}{2}\ln\frac{\eta(t)+\delta}{1-\eta(t)}, e(0) \geq 0 \\ \frac{1}{2}\ln\frac{1+\eta(t)}{\delta-\eta(t)}, e(0) < 0 \end{cases} \tag{11}$$

where $\eta(t) = e(t)/\sigma(t)$.

3.2. Fixed-Time Sliding Mode Controller Design

This study introduces a fixed-time SMC approach. Aiming at the velocity loop of PMLSMs, we use a Prescribed Performance Controller. In the current loop, a PI controller is implemented. The velocity tracking error is formally defined as:

$$e = v - v^* \tag{12}$$

where v^* is the reference speed. PPC is applied to speed tracking error. Take the derivative of ε to convert between the tracking error and the conversion error derivatives as follows:

$$\dot{\varepsilon} = \frac{\partial\psi^{-1}}{\partial\eta}\cdot\dot{\eta} = \frac{\partial\psi^{-1}}{\partial\eta}\cdot\frac{\dot{e}\sigma - e\dot{\sigma}}{\sigma^2} = m(\dot{e} - n) \tag{13}$$

where $m = \frac{\partial\psi^{-1}/\partial\eta}{\sigma}$, $n = \frac{\dot{\sigma}e}{\sigma}$.

Subsequently, the construction of the sliding mode surface is consisitent with the conversion error derived from the prescribed performance. The feedback controller is then crafted as a fixed-time sliding mode control, featuring an expressible sliding mode surface, formulated as:

$$s = \varepsilon + \int (\alpha_1 \mathrm{sig}(\varepsilon)^{\frac{2q_1-p_1}{q_1}} + \beta_1 \mathrm{sig}(\varepsilon)^{\frac{p_1}{q_1}})\mathrm{d}t \tag{14}$$

where $\mathrm{sig}^{\alpha}(x) = \mathrm{sign}(x)|x|^{\alpha}$, where $\alpha > 0$, $x \in R$, $\mathrm{sign}(x)$ is a standard symbolic function. The parameter α_1 and β_1 are positive real values, and p_1 and q_1 are positive odd integers satisfying $p_1 < q_1$. The introduction of the nonlinear function $\mathrm{sig}(x)$ into the sliding mode surface can make the slope of the sliding mode surface near the origin of the phase plane steeper and the convergence speed faster.

It is obtained by differentiating the sliding mode surface:

$$\begin{aligned} \dot{s} &= \dot{\varepsilon} + \alpha_1 \mathrm{sig}(\varepsilon)^{\frac{2q_1-p_1}{q_1}} + \beta_1 \mathrm{sig}(\varepsilon)^{\frac{p_1}{q_1}} \\ &= m(\dot{e} - n) + \alpha_1 \mathrm{sig}(\varepsilon)^{\frac{2q_1-p_1}{q_1}} + \beta_1 \mathrm{sig}(\varepsilon)^{\frac{p_1}{q_1}} \end{aligned} \tag{15}$$

The feedback control law is designed as follows:

$$\begin{aligned} i_q^* = \frac{1}{B_m}[&\dot{v}^* + n - A_m v + l\cdot\mathrm{sign}(s) \\ &- \frac{1}{m}(\alpha_1 \mathrm{sig}(\varepsilon)^{\frac{2q_1-p_1}{q_1}} + \beta_1 \mathrm{sig}(\varepsilon)^{\frac{p_1}{q_1}} + \alpha_2 \mathrm{sig}(s)^{\frac{2q_2-p_2}{q_2}} + \beta_2 \mathrm{sig}(s)^{\frac{p_2}{q_2}})] \end{aligned} \tag{16}$$

where the parameter α_2 and β_2 are positive real values, and p_2 and q_2 are positive odd integers satisfying $p_2 < q_2$.

Choose the Lyapunov function as

$$V = |s| \tag{17}$$

Taking the derivative of V yields

$$\begin{aligned}
\dot{V} &= sign(s) \cdot \dot{s} \\
&= sign(s) \cdot [m(A_m v + B_m i_q + D - \dot{v}^* - n) + \alpha_1 sig(\varepsilon)^{\frac{2q_1-p_1}{q_1}} + \beta_1 sig(\varepsilon)^{\frac{p_1}{q_1}}] \\
&= sign(s) \cdot [-\alpha_2 sig(s)^{\frac{2q_2-p_2}{q_2}} - \beta_2 sig(\varepsilon)^{\frac{p_2}{q_2}} - m(lsign(\varepsilon) - D)] \\
&= -\alpha_2 |s|^{\frac{2q_2-p_2}{q_2}} - \beta_2 |s|^{\frac{p_2}{q_2}} - m(l - D\text{sign}(s))
\end{aligned} \tag{18}$$

According to the assumption 2, we know that $|D| \leq l$, and $m > 0$.

$$\begin{aligned}
\dot{V} &= -\alpha_2 V^{\frac{2q_2-p_2}{q_2}} - \beta_2 V^{\frac{p_2}{q_2}} - m(l - D\text{sign}(s)) \\
&\leq -\alpha_2 V^{\frac{2q_2-p_2}{q_2}} - \beta_2 V^{\frac{p_2}{q_2}}
\end{aligned} \tag{19}$$

According to the lemma, we know that the control method proposed in this paper will converge in a fixed time and its convergence time is:

$$T \leq \left(\frac{1}{\alpha_2} + \frac{1}{\beta_2}\right) \frac{q_2}{q_2 - p_2} \tag{20}$$

From Equation (21), it is easy to see that larger α_2, β_2 and smaller values of p_2/q_2 will result in smaller convergence times for the system. It is important to note that a smaller convergence time does not necessarily mean better performance, as it requires a larger initial control input signal which may be difficult to achieve in practical control systems. The block diagram of PMLSMs control system are shown in Figure 1.

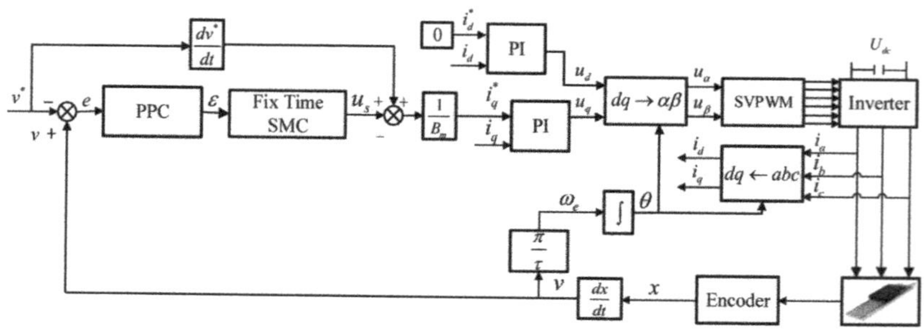

Figure 1. Block diagram of PMLSM control system using PPC-FTSMC strategy.

4. Simulations

To validate the efficacy of the designed PPC-FTSMC controller, this study employs MATLAB/Simulink software to construct a simulation model for the PMLSM drive control system. A comparative analysis between the PPC-FTSMC controller, FTSMC controller, and the conventional PI controller is conducted. The speed tracking response of the PMLSM under lumped disturbances are examined in the context of both control strategies. Nominal parameters for simulating the PMLSM are presented in Table 1.

Table 1. Parameters of the PMLSM in simulations

Parameters	Variable Name	Value
Stator resistance [Ω]	R_s	0.045
Stator Inductance [mH]	L_s	1.15
Mover mass [kg]	M	600
Friction coefficient [N·s/m]	B_v	0.5
pole-pitch [mm]	τ	200
permanent magnet flux linkage [Wb]	$flux$	0.145
number of pole pairs	n	2

To achieve superior control performance characterized by high tracking accuracy and fast convergence of state variables, a series of repeated simulation tests were conducted to identify the control parameters with good performance for the speed loops in the three controllers. The following settings were determined:

PI: $k_{pv} = 1850$, $k_{iv} = 19750$;

FTSMC : $p_1 = p_2 = 7, q_1 = q_2 = 9, \alpha_1 = \beta_1 = 30, \alpha_2 = \beta_2 = 350.$

PPC-FTSMC :The parameters of the sliding mode surface are consistent with FTSMC. The prescribed performance function is set as: $\sigma(t) = 0.1\text{e}^{-20t} + 0.01$ and $\delta = 1$.

For fair comparison, the current loops of the control strategies are PI controllers with the same parameters, and the proportional integral parameter value is set to $k_{pc} = 1.725$, $k_{ic} = 67.5$. The initial lumped disturbance D of the system is set to 2000 N. When $t = 2$ s, set the lumped disturbance D to increase from 2000 to 6500 N. In addition, the current i_q^* limiting value of the two controllers is set to ±1000 A, and the amplitude u_d^* and u_q^* limiting value of the voltage sum is set to ±1500 V. The simulation sampling time is set to 10 s. In order to display the superiority of the controller, the speed error datas of the motor are collected and analyzed quantitatively.

In order to evaluate the tracking performance of the designed controller under different operating conditions, the simulation and comparative analysis are carried out in this chapter.

Case 1: Set the following reference speed curve :

$$v^*(m/s) = \begin{cases} 4t, 0 \leq \text{t} \leq 1 \\ 4, 1 < \text{t} < 9 \\ 40 - 4t, 9 \leq \text{t} \leq 10 \end{cases} \tag{21}$$

The speed response, three-phase current, and electromagnetic thrust presented in Figures 2 and 3. As we can see, the three-phase current waveform of the three methods are relatively stable, but the current waveform of PI will jitter when the motor changes from accelerated operation to uniform operation at 1 s. And at 2 s, when the lumped disturbance changes, the current waveform of PI will peak. But the current waveform of FTSMC and PPC-FTSMC can remain stable throughout the motor operation. In terms of electromagnetic thrust, the thrust waveform of PI is overblown by a relatively obvious amount of about 20% in the starting state of the motor. In addition, when the machine acceleration changes and the lumped disturbance changes, the thrust waveform of the motor will be overblown, and the adjustment time is about 0.3 s. When adopt FTSMC, although the motor thrust waveform is not overshooting, the adjustment time is slightly longer than that of PPC-FTSMC. Throughout the entire control process, large electromagnetic thrust pulsation will cause motor jitter. Under the influence of the PPC-FTSMC, the three-phase current and electromagnetic thrust response speed of the motor are faster than those of the other controller, enabling quicker attainment of the steady state value. Furthermore, when the motor encounters abrupt lumped disturbances, the PPC-FTSMC promptly stabilizes the motor's electromagnetic thrust, diminishes the magnitude of speed variations, and reduces jitter.

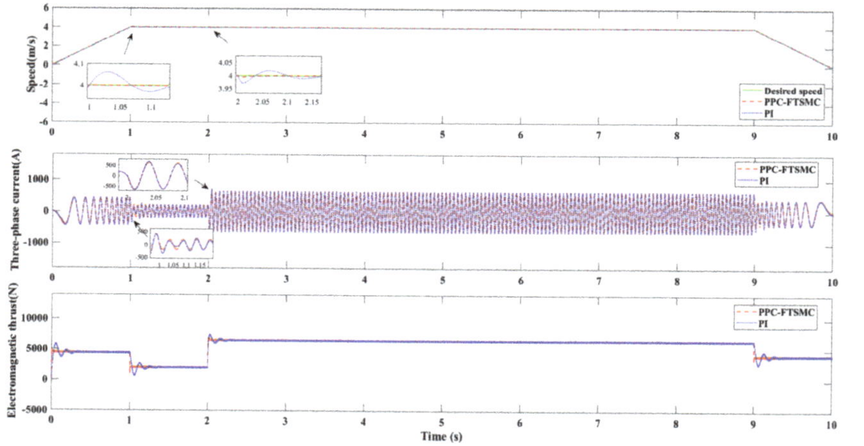

Figure 2. Simulation performance of speed response, three-phase current, and electromagnetic thrust compared with PI in case 1.

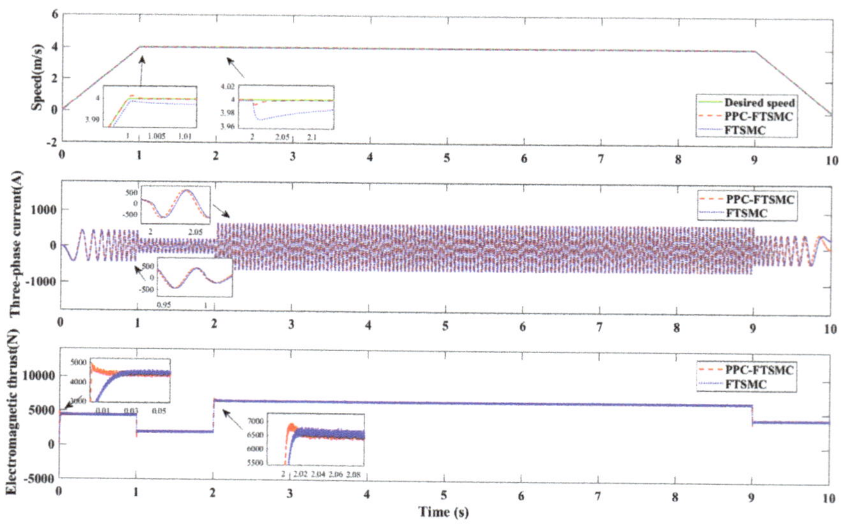

Figure 3. Simulation performance of speed response, three-phase current, and electromagnetic thrust compared with FTSMC in case 1.

Further, the tracking performance of the motor is analyzed by Figure 4, it can be seen that the PI controller fails to effectively track the specified speed signal and the overshoot controlled is undoubtedly the largest in the start stage of the motor. It takes approximately 0.4 s for the system to attain the desired speed and achieve a steady state. Although FTSMC is no overshoot, its response time is the longest, about 2 s, it proves that its speed response performance is the worst. PPC-FTSMC is used for the fastest dynamic response, with a response time of about 0.01 s and the smallest error significantly better than the other two methods. At 2 s, the system lumped disturbance changes, the motor speed tracking error of PPC-SMC can be effectively limited to the set 0.01 m/s, the adjustment time is about 0.02 s, while the maximum tracking error of PI control is 0.02 m/s, and the maximum tracking error of FTSMC control is 0.03 m/s.

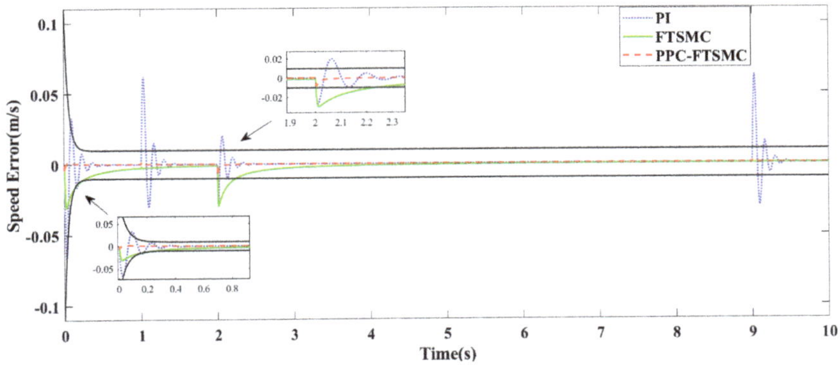

Figure 4. Speed tracking error in case 1.

Case 2: Set the reference speed curve to a sinusoidal signal, its frequency is 2 rad/s and amplitude is 5 m/s.

As shown in Figures 5 and 6, in terms of three-phase current, the three-phase current waveform of the three methods is relatively stable, but the three-phase current of PPC-FTSMC is more stable than the other two methods. When the motor is started, the thrust waveform will have a period of continuous 4 s oscillation in PI controller. Although the thrust of FTSMC is not overshot, it also has a response time lasting 0.02 s. The response time of thrust waveform of the motor system using PPC-FTSMC is the shortest, only 0.01 s, compared with other two algorithms. Similarly, when the load is disturbed for 2 s, PPC-FTSMC still has the fastest response time. It can be proved that when the whole system encounters lumped disturbance changes, the motor system using PPC-FTSMC can stabilize the electromagnetic thrust of the motor faster, reduce the amplitude of speed change, and reduce jitter. The speed signal of the motor is further analyzed according to Figure 7, when using PI control, the motor system cannot track the reference signal in time. When the speed signal reaches the amplitude, there will be a large speed tracking error of 0.01 m/s. In the motor system using FTSMC and PPC-FTSMC, the speed tracking error will approach 0, that is, the motor speed will eventually approach the given speed signal. The difference is that the dynamic adjustment performance of the motor system using FTSMC is poor, in the initial operation stage of the motor and when the load disturbance changes, the adjustment time is about 2 s, longer than that of PPC-FTSMC. In the whole operation process of the motor, whether in the initial stage or when the load disturbance changes, the speed tracking error of PPC-FTSMC can be kept within the limited 0.01 m/s.

In order to show the superiority of PPC-FTSMC controller, quantitative analysis of the motor's speed error data is conducted. The three error indicators listed in this article include maximum error (MAX), average absolute error (AVG), and root-mean-square error (RMS). The specific formula are as follows:

$$\text{MAXerror} = \max_{i} \sqrt{\varepsilon(i)^2} \tag{22}$$

$$\text{AVGerror} = \sum\nolimits_{i=1}^{N} \frac{|\varepsilon(i)|}{N} \tag{23}$$

$$\text{RMSerror} = \sqrt{\sum\nolimits_{\mathrm{i}=1}^{\mathrm{N}} \frac{|\varepsilon^2(\mathrm{i})|}{\mathrm{N}}} \tag{24}$$

The speed error indexes of three control strategies under two simulation conditions are given in Table 2. In terms of the error value, the speed tracking error of the motor system using PPC-FTSMC is much smaller than that of the other two methods. As can be seen from Table 2, the motor error value of PPC-FTSMC is less than 0.01 m/s given in the prescribed performance function. It is proved that the motor system with PPC has excellent performance in speed tracking control.

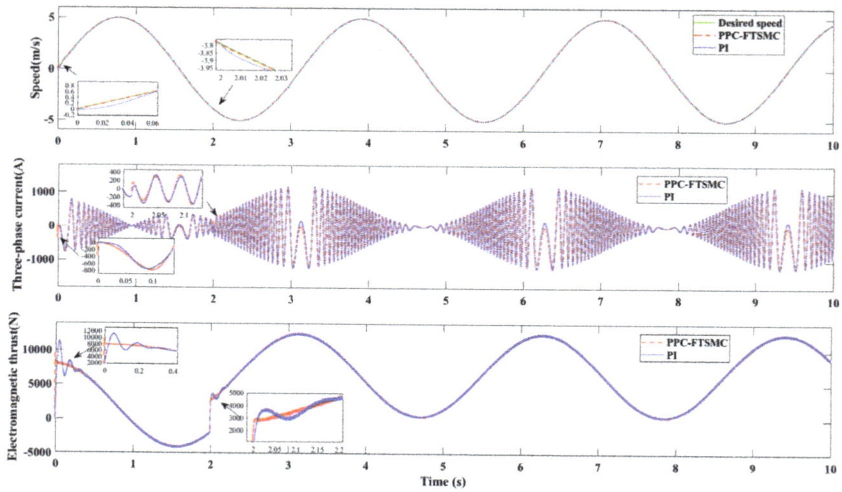

Figure 5. Simulation performance of speed response, three-phase current, and electromagnetic thrust compared with PI in case 2.

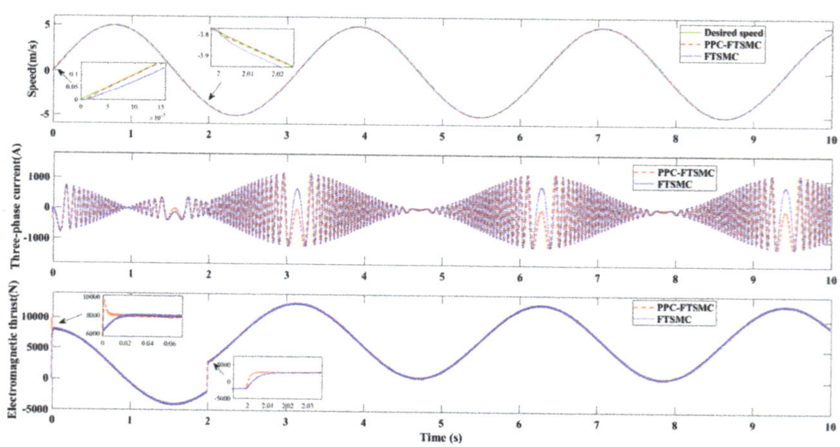

Figure 6. Simulation performance of speed response, three-phase current, and electromagnetic thrust compared with FTSMC in case 2.

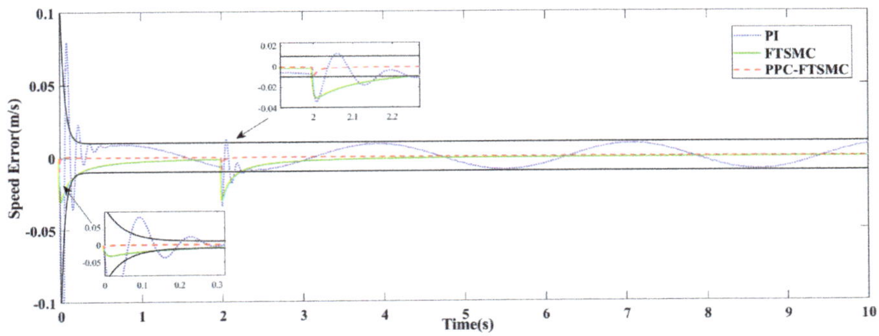

Figure 7. Speed tracking error in case 2.

Table 2. The speed errors comparisons using different methods.

Error	Case 1			Case 2		
	PPC-FTSMC	FTSMC	PI	PPC-FTSMC	FTSMC	PI
Maximum Error (m/s)	5.1×10^{-3}	3.1×10^{-2}	6.7×10^{-2}	9×10^{-3}	3.05×10^{-2}	1.59×10^{-1}
Average Absolute Error (m/s)	2×10^{-4}	8.3×10^{-3}	5.5×10^{-3}	2×10^{-4}	8.3×10^{-3}	1.83×10^{-2}
Root Mean Squared Error (m/s)	4×10^{-4}	1.07×10^{-2}	1.38×10^{-2}	5×10^{-4}	1.07×10^{-2}	3.4×10^{-2}

5. Conclusions

This paper introduces a fixed-time sliding mode controller with prescribed performance strategy, which is used for the control of a permanent magnet linear synchronous motor (PMLSM). Fixed-time sliding mode control is incorporated into the design of the speed control loop for the PMLSM, improving the system control performance , and achieving sliding mode controller stability in a fixed time, and accelerating the motor's speed error convergence. Furthermore, the prescribed performance function is applied to transform the constrained speed-tracking error into an unconstrained conversion error. Subsequently, a control law is formulated to limit the error within predefined bounds. This ensures that the system's dynamic characteristics, steady-state error, and other crucial performance indicators satisfy prescribed requirements. The efficacy of the proposed method is validated through simulation experiments. A comparative analysis is conducted against a PI controller and FTSMC, revealing that the proposed approach exhibits satisfactory control performance and induces smaller speed fluctuations. It is worth further considering that external disturbance will inevitably affect the operation of the PMLSM system. Although the prescribed performance control can limit the speed error within predefined bounds, its anti-interference effect is still not ideal. In follow-up work to this paper, a suitable disturbance observer will be design to improve anti-interference capability.

Author Contributions: Conceptualization, G.H. and C.Y.; methodology, G.H. and Q.S.; software, C.Y; validation, W.Y.; formal analysis, G.H.; investigation, C.Y.; data curation, Y.W.; writing—original draft preparation, G.H.; writing—review and editing, G.H.; supervision, W.Y.; project administration, Q.S. All authors have read and agreed to the published version of the manuscript.

Funding: This research received no external funding.

Data Availability Statement: The data and materials used to support the results of this study cannot be obtained due to privacy reasons.

Conflicts of Interest: The authors declare no conflict of interest.

References

1. Yin, X.; She, J.; Liu, Z.; Wu, M.; Kaynak, O. Chaos suppression in speed control for permanent-magnet-synchronous-motor drive system. *J. Frankl. Inst.* **2020**, *357*, 13283–13303. [CrossRef]
2. Cheng, S.; Yu, J.; Zhao, L.; Ma, Y. Adaptive fuzzy control for permanent magnet synchronous motors considering input saturation in electric vehicle stochastic drive systems. *J. Frankl. Inst.* **2020**, *357*, 8473–8490. [CrossRef]
3. Zhang, H.; Kou, B.; Jin, Y.; Zhang, H. Investigation of auxiliary poles optimal design on reduction of end effect detent force for PMLSM with typical slot–pole combinations. *IEEE Trans. Magn.* **2015**, *51*, 1–4.
4. Wang, M.; Yang, R.; Zhang, C.; Cao, J.; Li, L. Inner loop design for PMLSM drives with thrust ripple compensation and high-performance current control. *IEEE Trans. Ind. Electron.* **2018**, *65*, 9905–9915. [CrossRef]
5. Shi, Z.; Wang, Y.; Ji, Z. Bias compensation based partially coupled recursive least squares identification algorithm with forgetting factors for MIMO systems: Application to PMSMs. *J. Frankl. Inst.* **2016**, *353*, 3057–3077. [CrossRef]
6. Lin, F.J.; Hwang, J.C.; Chou, P.H.; Hung, Y.C. FPGA-based intelligent-complementary sliding-mode control for PMLSM servo-drive system. *IEEE Trans. Power Electron.* **2010**, *25*, 2573–2587. [CrossRef]
7. Cho, K.; Kim, J.; Choi, S.B.; Oh, S. A High-Precision Motion Control Based on a Periodic Adaptive Disturbance Observer in a PMLSM. *IEEE/ASME Trans. Mechatron.* **2015**, *20*, 2158–2171. [CrossRef]
8. Kommuri, S.K.; Park, Y.; Lee, S.B. Online Compensation of Mechanical Load Defects With Composite Control in PMSM Drives. *IEEE/ASME Trans. Mechatron.* **2021**, *26*, 1392–1400. [CrossRef]
9. Wang, M.; Li, L.; Pan, D.; Tang, Y.; Guo, Q. High-Bandwidth and Strong Robust Current Regulation for PMLSM Drives Considering Thrust Ripple. *IEEE Trans. Power Electron.* **2016**, *31*, 6646–6657. [CrossRef]
10. Wang, J.; Li, D. Application of fuzzy PID control in PMLSM servo drive system. In Proceedings of the 2015 IEEE International Conference on Mechatronics and Automation (ICMA), Beijing, China, 2–5 August 2015; IEEE: Piscataway, NJ, USA, 2015; pp. 6–10.
11. Chen, S.Y.; Liu, T.S. Intelligent tracking control of a PMLSM using self-evolving probabilistic fuzzy neural network. *IET Electr. Power Appl.* **2017**, *11*, 1043–1054. [CrossRef]
12. Wang, M.S.; Tsai, T.M. Sliding mode and neural network control of sensorless PMSM controlled system for power consumption and performance improvement. *Energies* **2017**, *10*, 1780. [CrossRef]
13. Jon, R.; Wang, Z.; Luo, C.; Jong, M. Adaptive robust speed control based on recurrent elman neural network for sensorless PMSM servo drives. *Neurocomputing* **2017**, *227*, 131–141. [CrossRef]
14. Lu, E.; Li, W.; Yang, X.; Liu, Y. Anti-disturbance speed control of low-speed high-torque PMSM based on second-order non-singular terminal sliding mode load observer. *ISA Trans.* **2019**, *88*, 142–152. [CrossRef] [PubMed]
15. Feng, Y.; Yu, X.; Man, Z. Non-singular terminal sliding mode control of rigid manipulators. *Automatica* **2002**, *38*, 2159–2167. [CrossRef]
16. Komurcugil, H. Non-singular terminal sliding-mode control of DC–DC buck converters. *Control Eng. Pract.* **2013**, *21*, 321–332. [CrossRef]
17. Moulay, E.; Léchappé, V.; Bernuau, E.; Defoort, M.; Plestan, F. Fixed-time sliding mode control with mismatched disturbances. *Automatica* **2022**, *136*, 110009. [CrossRef]
18. Hu, Y.; Yan, H.; Zhang, H.; Wang, M.; Zeng, L. Robust adaptive fixed-time sliding-mode control for uncertain robotic systems with input saturation. *IEEE Trans. Cybern.* **2022**, *53*, 2636–2646. [CrossRef] [PubMed]
19. Yu, L.; He, G.; Wang, X.; Zhao, S. Robust fixed-time sliding mode attitude control of tilt trirotor UAV in helicopter mode. *IEEE Trans. Ind. Electron.* **2021**, *69*, 10322–10332. [CrossRef]
20. Zamani, H.; Khandani, K.; Majd, V.J. Fixed-time sliding-mode distributed consensus and formation control of disturbed fractional-order multi-agent systems. *ISA Trans.* **2023**, *138*, 37–48. [CrossRef]
21. Pan, H.; Zhang, G.; Ouyang, H.; Mei, L. Novel fixed-time nonsingular fast terminal sliding mode control for second-order uncertain systems based on adaptive disturbance observer. *IEEE Access* **2020**, *8*, 126615–126627. [CrossRef]
22. Wang, L.; Du, H.; Zhang, W.; Wu, D.; Zhu, W. Implementation of integral fixed-time sliding mode controller for speed regulation of PMSM servo system. *Nonlinear Dyn.* **2020**, *102*, 185–196. [CrossRef]
23. Dai, Y.; Ni, S.; Xu, D.; Zhang, L.; Yan, X.G. Disturbance-observer based prescribed-performance fuzzy sliding mode control for PMSM in electric vehicles. *Eng. Appl. Artif. Intell.* **2021**, *104*, 104361. [CrossRef]
24. Song, Y.; Xia, Y.; Wang, J.; Li, J.; Wang, C.; Han, Y.; Xiao, K. Barrier Lyapunov function-based adaptive prescribed performance control of the PMSM used in robots with full-state and input constraints. *J. Vib. Control* **2023**, *29*, 1400–1416. [CrossRef]
25. Bu, X.; Wu, X.; Huang, J.; Wei, D. Robust estimation-free prescribed performance back-stepping control of air-breathing hypersonic vehicles without affine models. *Int. J. Control* **2016**, *89*, 2185–2200. [CrossRef]
26. Qin, H.; Wu, Z.; Sun, Y.; Chen, H. Disturbance-observer-based prescribed performance fault-tolerant trajectory tracking control for ocean bottom flying node. *IEEE Access* **2019**, *7*, 49004–49013. [CrossRef]
27. Wang, X.; Wang, X.; Wang, Z.; Xiao, X.; Li, S. Composite finite-time control for PMSM with prescribed performance using disturbance compensation technique. *Control Eng. Pract.* **2023**, *141*, 105677. [CrossRef]

28. Ding, B.; Xu, D.; Jiang, B.; Shi, P.; Yang, W. Disturbance-observer-based terminal sliding mode control for linear traction system with prescribed performance. *IEEE Trans. Transp. Electrif.* **2020**, *7*, 649–658. [CrossRef]
29. Bechlioulis, C.P.; Rovithakis, G.A. Robust adaptive control of feedback linearizable MIMO nonlinear systems with prescribed performance. *IEEE Trans. Autom. Control* **2008**, *53*, 2090–2099. [CrossRef]

MDPI
St. Alban-Anlage 66
4052 Basel
Switzerland
www.mdpi.com

Energies Editorial Office
E-mail: energies@mdpi.com
www.mdpi.com/journal/energies

www.ingramcontent.com/pod-product-compliance
Lightning Source LLC
Chambersburg PA
CBHW061603210926
51850CB00009B/2198
* 9 7 8 3 7 2 5 8 1 1 6 9 4 *